Lecture Notes in Economics
and Mathematical Systems
575

Rolf Hellermann

Capacity Options for Revenue Management

Theory and Applications
in the Air Cargo Industry

With 65 Figures
and 17 Tables

 Springer

Dr. Rolf Hellermann
Ulmenweg 4
32760 Detmold, Germany
E-mail: rolf.hellermann@whu.edu

 ISBN-10 3-540-34419-5   Springer Berlin Heidelberg New York
ISBN-13 978-3-540-34419-3   Springer Berlin Heidelberg New York

Springer is a part of Springer Science+Business Media
springer.com

© Springer-Verlag Berlin Heidelberg 2006
Printed in Germany

Typesetting: Camera ready by author
Cover: Erich Kirchner, Heidelberg
Production: LE-TEX, Jelonek, Schmidt & Vöckler GbR, Leipzig

SPIN 11759386   Printed on acid-free paper – 88/3100/YL – 5 4 3 2 1 0

Foreword

Arguably the central problem in Operations Research and Management Science (OR/MS) addressed by e-business is better coordination of supply and demand, including price discovery and reduction of transaction costs of buyer-seller interactions. In capital-intensive industries like air cargo, the out-of-pocket costs of excess capacity and the opportunity costs of underutilized capacity have been important factors driving the growth of exchanges for improving demand and supply coordination through e-business platforms.

Hellermann addresses in his dissertation one of the most interesting aspects of this evolution for OR/MS, the parallel development of long-term and short-term markets for capacity and output, accompanied by a range of option and fixed-commitment (i.e., forward) contracts as the basic mechanisms supporting transactions. This has been a fascinating topic for OR/MS research because it builds on the powerful framework of real options, while connecting directly to key operations decisions (capacity planning, network design, staffing, routing, maintenance, and so forth) of the equipment and technologies whose output is the focus of contracts. From the perspective of practice, the integrated use of these Internet-based contracting mechanisms, as facilitated by new B2B exchanges, represents an opportunity for further improving supply chain performance and capital asset productivity.

As Hellermann notes, a central feature of B2B for capital-intensive industries is that contracting needs to take place well in advance of actual delivery. Failure to do so for a non-scalable technology is a recipe for last-minute confusion and huge excess costs, e.g., offloading in the case of air cargo. This has given rise to a general recognition that most of the firm's output in such services industries should be contracted for well in advance. However, there is still a very important role for short-term fine-tuning of capacity and output to contract for, say, the last 10% of a firm's output or a customer's requirements. Doing so requires a conceptual framework, congenial to e-business, that allows contracting to take place at various points of time, constrained by various commitment and delivery options and flexibilities, and mediated by electronic markets where these are feasible. What Hellermann does here is to characterize the form of options-based instruments required to support this evolution. His results characterize the optimal form of options on capacity and related forward contracts.

In particular, in Chap. 2, Hellermann describes the practice of capacity reservation and dynamic pricing at Lufthansa Cargo AG. In this thesis, the interaction between freight forwarders and carriers is the main focus of the analysis. Shortcomings of existing contracts for the advance sale of capacity in this special services industry are being discussed. In the literature review, see Chap. 3, an extensive overview of flexible contracts for capacity contracting is presented. In Chap. 4, an innovative option pricing model for capacity reservation is proposed. The model accounts for risk in both demand and market price. Chapter 5 contains a comparative statics analysis of the model where fixed-commitment and option-type contracts are being benchmarked. In addition, the Pareto- or win-win efficiency of such option contracts is illustrated (for a wide range of parameter settings). Chapter 6 captures the case of overbooking which is prevalent in the industry. Chapter 7 utilizes sample data obtained from Lufthansa Cargo AG to test the applicability and impact of option-type contracts. The thesis closes with managerial implications, see Chap. 8.

This dissertation represents a solid piece of mature scholarship. The analysis is concise and splendidly readable. The insights for both theory and practice are trenchant. The findings are well connected to the literature of operations and finance, as well as to the broader arena of economics and market efficiency. This research provides a solid platform for further developments and for launching a research career in business studies. In short, this dissertation achieves outstanding marks on everything we consider important for a doctoral dissertation.

We acknowledge financial support through the ADVENTURES (Analysis of Dynamic Ventures Using Real-options in Services) grant under the project number 01HG9992/5, provided by the German Ministry of Education and Research in Bonn, Germany.

Arnd Huchzermeier   *Stefan Spinler*
Vallendar, April 2006   Leipzig, April 2006

Preface

This work was submitted as doctoral thesis at WHU, Otto Beisheim School of Management in October 2004. It summarizes research that I conducted at the Department of Production Management, chaired by Professor Dr. Arnd Huchzermeier, from 2001 to 2004.

I would like to thank Arnd Huchzermeier for the guidance he provided throughout this project and for the inspiring environment with international exposure he creates at his department. Professor Dr. Manfred Krafft, Director of the Institute of Marketing at the University of Münster, kindly agreed to act as co-advisor for my thesis. I am grateful to Dr. Stefan Spinler, Assistant Professor at the Department of Production Management, for always being available for discussions and for being open to my questions with regard to the formulation of the mathematical model.

Furthermore, I am indebted to Felix Keck, at that time Vice President Margin Management at Lufthansa Cargo AG, for providing the professional insights into the air cargo industry that otherwise would have been difficult to obtain. Lufthansa Cargo AG kindly provided the data for the application case study presented in Chap. 7.

The more difficult moments of the dissertation process were eased by my colleagues at the Department of Production Management who made the time spent at WHU in Vallendar a period I will always gladly recall. My "roommate" Rainer Brosch was a great support and sparring partner.

The most thankful I am to my parents, who from the very beginning opened up the way for the education I received over the years and always backed my plans without reservation. To Natalie, my fiancée, I would like to be forever as supportive as she was to me throughout this work.

Detmold, April 2006   *Rolf Hellermann*

Contents

List of Abbreviations, Variables, and Functions

*	(indicates optimality)	CPA	Capacity Purchasing Agreement
~	(indicates uncertainty)	c_u	Underage cost
3PL	Third-party logistics	$\tilde{D}_C$	Contract market demand
4PL	Fourth-party logistics	$\tilde{D}_S$	Spot market demand
A	(subscript:) Asset provider	$\tilde{E}$	Number of called on reservations
a	Maximum contract market size (ordinate intercept of contract market demand function)	$\mathcal{E}[\cdot]$	Expectation operator
		$\tilde{e}$	Stochastic error term in contract market demand function
ACC	Available capacity for contract		
B	Booking level	e.g.	exempli gratia (for example)
b	Price responsiveness of contract market demand (slope of contract market demand function)	Eq.	Equation
		et al.	et alii (and others)
		f	Fixed cost
B2B	Business-to-business	$F_i(\cdot)$	Cumulative distribution function of demand i
C	(subscript:) Contract market		
		$f_i(\cdot)$	Distribution density function of demand i
c	Variable reservation cost in contract market		
cf.	confer (compare)	f., ff.	following
c_o	Overage cost		

$\bar{F}_i(\cdot)$ Primitive of the cumulative distribution function of demand i

FS   (superscript:) Scenario with fixed-commitment contract and spot market sales

FTK   Freight-tonne kilometers

G   (subscript:) Integrated firm

Γ Total expected overbooking cost

g Offload cost

GCA   Guaranteed Capacity Agreement

$G(\cdot)$ Cumulative distribution function of spot price

$g(\cdot)$ Distribution density function of spot price

GDP   Gross domestic product

GF-X   Global Freight Exchange

η Price elasticity of demand

I   (subscript:) Intermediary

i.e.   id est (that is to say)

IATA   International Air Transport Association

ibid.   ibidem (the same)

κ Expected capacity utilization

K Capacity

ℓ Lagrange multiplier

λ Intermediary's markup

$\mathcal{L}$ Lagrangian

$L(\cdot)$ Standard normal loss function

LCAG   Lufthansa Cargo AG

$\tilde{M}$ Spot market sales

μ_i Mean (expected value) of i

N Number of reservations

n/a   not applicable

O&D   Origin and Destination

OS   (superscript:) Scenario with capacity-option contract and spot market sales

Π Expected profit

P Profit

p Price paid by shippers to forwarder

p., pp.   page(s)

PARM   Perishable asset revenue management

$\Phi(\cdot)$ Cumulative standard normal distribution function

$\varphi(\cdot)$ Standard normal distribution density function

$\tilde{Q}$ Offload quantity

QF   Quantity flexibility

r Reservation fee

R&D   Research and development

$\rho_{i,j}$ Coefficient of correlation between i and j

RM   Revenue management

RTK   Revenue-tonne kilometers

S   (subscript:) Spot market

S   (superscript:) Scenario with spot market sales only

$\tilde{s}$ Spot price

SARS   Severe acute respiratory syndrome

SCM   Supply chain management

σ_i Standard deviation of i

SRS   Standard Rate Sheet

t Variable cost in spot market

ϑ_i	Coefficient of variation of i
TKO	Tonne-kilometers offered
TKT	Tonne-kilometers taken
U.S.	United States
v	Variable execution cost in contract market
w	Capacity price in the fixed-commitment contract
x	Execution fee
z	Standardized random variable

1

Introduction

In today's world economy that is marked by increasing trade and volatility, air cargo acts as a facilitator exhibiting steady increases with an annual growth rate of more than 7% since the 1970s (Boeing 2004, p. 11). At the same time, the trend among manufacturing companies to concentrate on core competencies and outsource non-core activities continues unbroken. Especially the responsibility for transportation services is more and more passed on to specialized forwarding and logistics companies, commonly referred to as third-party ("3PL") logistics providers (cf. Murphy and Poist 2000).

On the market for airfreight transportation, there are mainly two types of players facing each other. On the sell side, air cargo carriers offer capital-intensive capacity that must be filled in order to generate their required return on capital. The buy side is dominated by freight forwarding and logistics service companies who try to secure capacity access while pressing for favorable terms.

Sellers strive to assure capacity utilization and mitigate cash flow risk by engaging in advance sale of capacity via long-term contracts. Buyers acting as resellers (intermediaries) are reluctant to commit because they are facing uncertain demand. They expect compensation for the loss of flexibility associated with long-term contracts in form of price breaks. The need for flexibility is even higher since overcapacity in the industry increases the chance that cheap capacity becomes available in the spot market. Air cargo carriers face the problem of designing and especially pricing contracts for advance sale of capacity that incorporate the desired flexibility.

The predominant type of long-term capacity agreement between air cargo carriers and forwarding companies today is a fixed-commitment (forward) contract (Pompeo and Sapountzis 2002), reserving a certain amount of capacity at an agreed-upon rate – payable after capacity usage – on a certain flight for the shipments delivered by the forwarding companies' customers.

Though only some of these contracts actually exhibit a cancellation clause, carriers rarely can enforce the terms of contract vis-à-vis their most important customers, leaving in effect the carrier with the entire utilization risk while giving the forwarder a free call option on capacity.

A contract type currently considered by airline managers is the capacity-option contract, which has, in a different context, been proposed in the recent supply chain management literature (Barnes-Schuster et al. 2002; Spinler 2003, cf.). A forwarding company that signs such an options contract would acquire the right but not the obligation to use the agreed-upon capacity, with a per-unit reservation fee payable ex-ante on signing the contract and a per-unit execution fee payable if capacity is eventually used. By setting reservation and execution fee appropriately, the carrier could adequately price the flexibility offered to the forwarder and potentially ease contract enforcement.

The subject of this thesis is the evaluation of option contracts' suitability to provide for the desired flexibility, the pricing of capacity through option contracts, and the valuation of the financial impact of capacity-option contracts as compared to fixed-commitment contracts. The analysis is conducted by means of an analytical, multi-variate optimization model under price and demand uncertainty. Through an application and feasibility study conducted on the basis of empirical data from a leading air cargo carrier, the applicability and potential impact of capacity-option contracts is demonstrated. Furthermore, it is shown how capacity-option contracts integrate into the context of air cargo revenue management.

The contributions of the thesis to the supply chain management literature are threefold and include

- the development of the capacity-option pricing model,
- the application case study that applies the model to a data set from a leading air cargo carrier, and
- demonstrating under which conditions capacity-option contracts are to be preferred over fixed-commitment contracts.

The key results established in the following chapters include that

- the seller is better off selling capacity options instead of fixed-commitment contracts except for certain market conditions;
- if the seller chooses to sell capacity options, this leads, under most circumstances, to a Pareto improvement, i.e., the buyer benefits, too (or is at least not made worse off);
- however, under rare circumstances, the seller's choice of a capacity-option contract makes the buyer worse off than a fixed-commitment contract, i.e., a Pareto improvement is not achieved;
- the improvement potential suggested by the model is confirmed by encouraging results from the application case study.

The structure of the subsequent chapters is as follows: Chapter 2 introduces into the research problem by giving an overview of the air cargo industry, its characteristics and current challenges. Through a case study on a world-class air cargo carrier, Lufthansa Cargo AG, the current state of capacity reservation and dynamic pricing in the industry is illustrated. Chapter 3 reviews the relevant literature, including the fields of advance sale of capacity, supply contracts, and revenue management.

Chapter 4 contains the formulation of the analytical model and the derivation of the optimal policies of capacity buyer and seller. The results of the model are presented in Chap. 5, including an illustration of the optimal policies and a comparative static analysis of the exogenous model parameters. Chapter 6 provides extensions and analyses of the model beyond the previously made assumptions, especially including an overbooking model. The insights from the application and feasibility study are presented in Chap. 7. Finally, Chap. 8 proposes managerial implications and concludes the work.

Capacity Agreements in the Air Cargo Industry

This chapter introduces the subject of capacity agreements in the air cargo industry. At first, an overview of the air cargo industry with its characteristics and current challenges is given. Then, the current state of capacity reservation and dynamic pricing is illustrated considering as example a major air cargo carrier, namely Lufthansa Cargo AG. The chapter concludes with the formulation of the research questions to be answered in subsequent chapters.

2.1 The Air Cargo Industry

The players in the market for air cargo transportation can be divided into three groups: asset providers, shippers, and intermediaries. In the following, the suppliers that offer airport-to-airport transport and operate physical assets (aircraft) that provide air cargo capacity are subsumed under the term *asset provider*. These are in the first place cargo-only carriers that operate freighter aircraft, passenger airlines that offer lower-deck freight capacity, and carriers offering both. Examples for the latter case include the freight subsidiaries of major airlines, e.g., Lufthansa Cargo AG, Air France Cargo, and Singapore Airlines Cargo.

The term *shippers* designates the airfreight senders. Only for a small part (approx. 5–10%, according to Althen et al. 2001, p. 424) of total airfreight volume, shippers send freight directly with asset providers (see Fig. 2.1). For the major part (approx. 90–95%) of volume, shippers leave it to *intermediaries* to organize and perform transportation (cf. Schneider 1993; Doganis 2002, p. 315). These intermediaries can be freight forwarding companies that operate trucks to cover the door-to-airport and airport-to-door sections of the airfreight transport. Freight forwarding companies that have extended their activities beyond simple (road) transportation to providing integrated

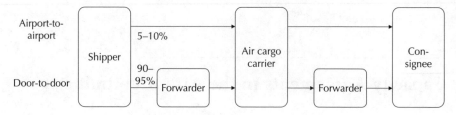

Fig. 2.1. The air cargo supply chain

logistics services that include, e.g., handling, storage, commissioning, and organization of transport chains are often referred to as third-party logistics (3PL) companies (cf. Skjoett-Larsen 2000; Herrmann et al. 1998b, p. 150).[1]

In the course of focusing activities on core competencies, many companies have outsourced their logistics requirements to freight forwarding companies and logistics service providers (cf. Murphy and Poist 2000). Typically, large shippers close agreements with such intermediaries about the terms and rates of service with regard to freight volume, weight, origin, and destination. For those sections of transport chains where the intermediary itself does not operate means of transport, e.g., sea and air, the intermediary purchases capacity from asset providers, e.g., shipping companies and airfreight carriers (cf. Lieb et al. 1993).

The so called "integrators", e.g., Federal Express and United Parcel Service, represent an exception to this business model. They provide door-to-door transport as an integrated service concept and own and operate all transport assets themselves. So far, these companies have focused on goods and packages up to 50 kilogram. Nevertheless they pose a challenge to non-integrated carriers who also try to attract business in this high-margin segment, but usually lack the selling proposition of a seamless door-to-door transport (cf. Doganis 2002, p. 312 ff.). Because of the absence of the necessity for capacity agreements that structure the relationship between asset provider and intermediary, this market segment is not further considered in the following.

Airfreight carriers have traditionally been anxious to maintain good business relationships to forwarders and logistics-service providers because these represent large aggregated volumes and control the direct contact with end customers for whom they usually decide on the actual carrier (cf. Weisskopf 1984, p. 172 ff.).

---

[1] A 4PL provider takes this concept one step further by not owning any transport asset or operating any part of the transportation network itself, but focusing on the organization and coordination of an entire supply chain (cf. Barde and Mueller 1999). For the following analysis, this distinction is of no further relevance; 3PLs and 4PLs are collectively referred to as logistics-service providers.

It is standard industry practice that airfreight carriers and intermediaries close capacity agreements by which intermediaries reserve or purchase airfreight capacity upfront and en bloc (cf. Herrmann et al. 1998a; Pompeo and Sapountzis 2002). An intermediary benefits from signing a capacity agreement because it secures capacity access if capacity is scarce and locks in prices. The incentive for the airfreight carrier to write capacity agreements is the reduction of capacity utilization risk since the agreements partially shift this risk to the contractual partner. (The motives for engaging in advance sale of capacity are discussed in greater detail in Sec. 3.1.1.) Capacity agreements, however, cannot be regarded in isolation but have to be seen within the context of an airlines product offering and its revenue management system.

2.1.1 Challenges in Air Cargo Transportation

Though closely related and often even sharing resources and equipment, the air cargo business differs from the passenger business. Especially with regard to network planning and capacity allocation, cargo carriers have more degrees of freedom and hence face additional complexity as compared to passenger airlines (cf. Kasilingam 1996, pp. 37 f.):

- Unlike passengers, cargo shippers do not book round trips. Thus cargo flows are unpaired and, even on a global level, are not necessarily balanced.
- Cargo is characterized by multiple dimensions (volume and weight); while each passenger can be assigned exactly one seat, cargo is characterized by weight, volume, and the number of container positions required aboard the aircraft. The load can be balanced and optimized by mixing shipments with different specific weights, i.e., volume-to-weight relations. Ideally, space can be sold twice, e.g., to one customer with voluminous, light cargo and another with heavy-weight, high-density cargo.
- While passengers purchase tickets for specific flights and routes, cargo airlines can transport goods flexibly with regard to time and route through their network, the only constraint being the promised time of availability at destination.
- On passenger aircraft, the capacity available for cargo is uncertain over the booking horizon; it depends on the number of passengers and the volume and weight of their baggage.

2.1.2 Market Dynamics

The range of products transported by airfreight has grown beyond documents and traditional air cargo goods (like electronics) to include fash-

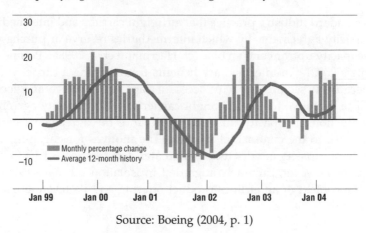

Source: Boeing (2004, p. 1)

Fig. 2.2. Demand volatility in the world air cargo market: Monthly percentage of change in volume over prior year.

ion goods, perishables, machinery components and spare parts, etc. This growth has been fueled by manufacturers more and more adopting just-in-time strategies and consumers more and more demanding international products (Shields 1998, p. 184). However, these practices have also added to volatility of demand in the world air cargo market. Fluctuations of demand by ± 15 to 20% within one year are not unusual (see Fig. 2.2).

The major driver behind these fluctuations is the global economy that drives world trade and thus the demand for airfreight transportation services. Fig. 2.3 shows the close link between the development of the growth rate of the world air cargo market and the world gross domestic product (GDP) over the past 20 years. Though the average annual growth rate of the world air cargo market, measured in revenue-tonne kilometers (RTK), amounted to 7.1% since 1970 (Boeing 2004, p. 11), it fluctuates widely and also exhibits phases of market contraction at the beginning of the 1990s and the current decade. Both growth and variability of the air cargo market are typically higher than the world GDP's. Economic cycles thus hit air cargo carriers in an amplified way (cf. Financial Times 2004b).

This poses a challenge for air cargo carriers to plan and adapt capacity accordingly. However, given the lumpy nature and capital intensity of airfreight capacity, capacity cannot easily be changed at short notice. This increases the relevance of risk sharing by advance sale of capacity and the general application of revenue management practices (see Sec. 2.2.3 and 3.3).

During the years of rising demand at the end of the 1990s, carriers have built up freighter capacity which now, after the economy and consequently the demand for air cargo has slowed down (see Fig. 2.2), results in the industry suffering from overcapacity (cf. Kay 2003) because carriers can adjust

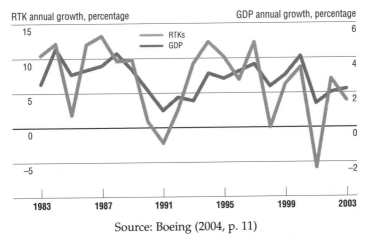

Source: Boeing (2004, p. 11)

Fig. 2.3. Historic growth of the global air cargo market: Market growth measured in revenue tonne-kilometers (RTK) is correlated with growth of the world gross domestic product (GDP).

physical capacity only in relatively large increments[2], determined by aircraft size. Pompeo and Sapountzis (2002, p. 92) observe that "[c]ompanies tend to order these [aircraft] simultaneously, when they think the world economy is set to grow. If, as can easily happen, they make a collective mistake about the cycle's timing, they might take delivery of new capacity just as demand drops."

Furthermore, since between 40–50% of global airfreight capacity is made up by belly capacity on passenger aircraft, airfreight capacity supply is partly driven by an unrelated market demand, i.e., demand for air passenger transport (Kadar and Larew 2004, p. 4). This and the above reasons make it generally difficult for airlines to match demand and supply.

Boeing (2004, p. 5) projects the air cargo market growth to continue with more than 6% annually over the years 2004–2023. The aircraft manufacturer also forecasts the world freighter aircraft fleet to grow from 1,766 in 2003 to more than 3,400 in 2023 (ibid., p. 96) and projects a trend to larger aircraft. In 2001, the average load capacity of a freighter amounted to 48 tons, but is projected to grow to up to 60 tons over the subsequent 20 years (Boeing 2002, p. 95), partly as a result of the future availability of larger aircraft like the Airbus A380.

A further component of the dynamics in the air cargo market is the development of yield over time. Yield refers to the average revenue per revenue-tonne kilometer (RTK). In general, yield was declining at –3.4% an-

---

[2] For example, adding one freighter to Lufthansa Cargo's fleet (see Sec. 2.2 in combination with the information from Table 2.1) increases total capacity by 4–5%.

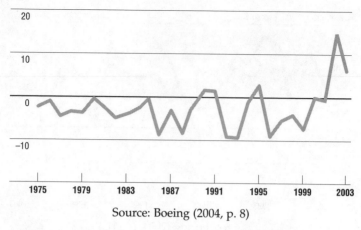

Source: Boeing (2004, p. 8)

Fig. 2.4. Annual percentage change of world airfreight yield

nually from 1985 to 2001, while the passenger yield, for comparison, was declining at –2.1% over the same time horizon (Coyne 2003; Boeing 2002, p. 8).
However, annual change of the yield growth rate has fluctuated between –
10% and +15% (see Fig. 2.4), with the increase observable for the years 2002
through 2003 being attributable to the U.S. West Coast Port Strike and SARS,
which produced a shortage of cargo capacity in certain major markets, to
worldwide economic recovery, and to fuel and security surcharges (Boeing
2004, p. 2).

2.2 Case Study: Capacity Reservation and Dynamic Pricing at Lufthansa Cargo AG[3]

Lufthansa Cargo AG (LCAG) is a wholly owned subsidiary of Deutsche
Lufthansa AG. LCAG owns and operates a fleet of 22 dedicated freighter
aircraft, namely 8 Boeing B747-200 and 14 MD-11F aircraft (Lufthansa Cargo
2004) with 100 and 85 tons load capacity, respectively. In addition, it markets any lower-deck capacity of Lufthansa's passenger aircraft not occupied
by passengers' baggage. LCAG purchases this space en bloc from the par-
ent company at transfer prices. With a 6.5% share of total market volume
(Sowinski 2002), the company is among the market leaders in the world international air cargo market (see Table 2.1). In the fiscal year 2003, revenues

---

[3] This section is partially based on and summarizes findings from Hellermann
and Huchzermeier (2002a, b). It has benefited greatly from ongoing discussions
with managers of Lufthansa Cargo AG. If not stated otherwise, the information
presented in this section is based on interviews and discussions with LCAG
management.

amounted to € 2.2 billion, resulting in earnings before income tax of € 16 million (Lufthansa Cargo 2004).[4]

2.2.1 Products

LCAG offers standardized products in all three major market segments of the global airfreight market. These include express products for which the company guarantees capacity availability (within certain weight restrictions) and reimbursed freight charges within the scope of a performance guarantee if it fails to deliver on time. For all products, the company promises availability at the airport of destination within an agreed-upon time frame. LCAG markets its products as "time-definite" (td) services named td.Pro, td.X, td.Flash, and td.SameDay[5]. Customers can combine products with standardized "packages and options" in the case of goods that require special services like perishables, live animals, or fragile products. For example, if a customer books LCAG's package "smooth/td" to transport semiconductor wafers, it includes making the aircraft take off and touch down at a lower angle than usual in order to preserve the highly fragile high-tech products.

2.2.2 Competition

LCAG faces various forms of competition: Other airlines are competitors in the airport-to-airport transport business. Some of them – like Lufthansa – operate and own cargo subsidiaries and/or freighter fleets, e.g., Air France Cargo, Korean Air, or Emirates SkyCargo. With three of its competitors – SAS Cargo, Singapore Airlines Cargo, and Japan Airlines Cargo –, LCAG has recently formed a strategic alliance, initially called *New Global Cargo* and later renamed *WOW*[6], that aims at enlarging the network of destinations the carriers offer to their clients (cf. Grin 1998, p. 83).

Competition also arises from airlines focusing exclusively on cargo transport, e.g., Cargolux, and even from airlines that operate passenger aircraft only (for an overview of competitive strategies in the air cargo industry, see

---

[4] Considering the company's meager current financial performance, LCAG announced a restructuring program in 2004, aimed at increasing productivity. The program includes a workforce reduction by 10%. The carrier is cited to expect to return to a 10% profit margin by 2006. In the first half of 2004, LCAG reported a loss of € 3 million on sales of € 1.2 billion, reflecting the negative impact of currency effects and increases of fuel price (Financial Times 2004a).

[5] The products differ with respect to promised time frame of availability at destination, shipment weight, and capacity guarantee.

[6] Besides the connotations related to its pronunciation, this letter sequence has no further acronymic meaning.

Table 2.1. World's leading international air cargo carriers by scheduled freight-tonne kilometers (FTK) performed in international service in 2000.

Int'l Rank	Airline	Int'l FTK (millions)	Share of carrier's total FTK (int'l)	World int'l market share	% of int'l FTKs on freighters
1	Lufthansa Cargo	7,069	99.7%	6.5%	64.7%
2	Korean Air Lines	6,357	98.4%	5.8%	82.9%
3	Singapore Airlines	6,020	100.0%	5.5%	54.9%
4	Air France	4,968	99.8%	4.6%	56.1%
5	British Airways	4,555	99.8%	4.2%	16.9%
6	Federal Express	4,456	41.2%	4.1%	100.0%
7	Japan Airlines	4,321	93.8%	4.0%	54.7%
8	China Airlines	4,136	100.0%	3.8%	90.0%
9	Cathay Pacific	4,108	100.0%	3.8%	53.1%
10	KLM	3,964	100.0%	3.6%	16.2%
11	EVA Air	3,558	100.0%	3.3%	90.0%
12	Cargolux	3,523	100.0%	3.2%	100.0%
13	United Airlines	2,777	75.2%	2.5%	10.3%
14	Asiana	2,607	100.0%	2.4%	50.0%
15	Northwest Airlines	2,409	74.5%	2.2%	57.2%
16	Martinair Holland	2,356	100.0%	2.2%	90.0%
17	Nippon Cargo Airlines	2,186	100.0%	2.0%	100.0%
18	United Parcel Service	2,174	34.4%	2.0%	100.0%
19	American Airlines	2,166	77.9%	2.0%	0.0%
20	Swissair/Swiss	1,930	99.9%	1.8%	7.3%
21	Malaysia Airlines	1,812	97.0%	1.7%	0.0%
22	Alitalia	1,734	99.5%	1.6%	52.6%
23	Thai Airways	1,678	98.0%	1.5%	6.9%
24	Qantas	1,531	94.6%	1.4%	3.9%
25	Delta Airlines	1,525	72.8%	1.4%	0.0%
	Top 25 scheduled int'l	83,947	85.7%	77.0%	56.6%
	Other IATA airlines	25,102	85.8%	23.0%	21.0%
	Total scheduled int'l	109,049	85.7%	100.0%	48.4%

Source: Sowinski (2002)

Table 2.2. World's leading international airfreight forwarders by cargo tonnage shipped in international service in 2000.

Int'l Rank	Freight forwarding company	Tons (thousands)	Market share (int'l)
1	Danzas (DHL)	1,835	11.6%
2	Panalpina	1,249	7.9%
3	Schenker (Stinnes)	701	4.4%
4	Kühne & Nagel	700	4.4%
5	Nippon Express	693	4.4%
6	BAX Global	655	4.2%
7	Emery/Menlo (CNF)	586	3.7%
8	Kintetsu	559	3.5%
9	Exel	525	3.3%
10	Expeditors	507	3.2%
11	EGL Global Logistics	380	2.4%
12	Hellmann	365	2.3%
13	Fritz (UPS)	348	2.2%
14	UTI	228	1.4%
15	SDV Group (Bollor)	224	1.4%
	Total top 15	9,556	60.6%
	All other competitors	6,214	39.4%
	Total international markets	15,770	100.0%

Source: Ott (2003)

Althen et al. 2001; Zhang and Zhang 2002 discuss the different impact of air cargo liberalization on all-cargo and mixed passenger/cargo carriers). As the latter tend to consider cargo a mere by-product (cf. Grin 1998, p. 78), some apply marginal pricing which has led to an erosion of cargo rates in the period since the 1950s and overcapacity on highly frequented passenger routes. In the express product sector, LCAG's major competitors are integrators like Federal Express and United Parcel Service. Both companies operate own freighter fleets.

While being LCAG's best and most important customers, forwarding companies are at the same time its hardest competitors with regard to direct customer contact. The top 15 forwarders are reported to control 60% of the tonnage moved in international air cargo (see Table 2.2). Most shippers deal primarily with and make payments to forwarders, not air carriers, significantly reducing the latter's share of revenue. LCAG tries to counter this trend by having established a *Business Partnership Program* with select forwarding companies, tying these closer to the airline. The program includes

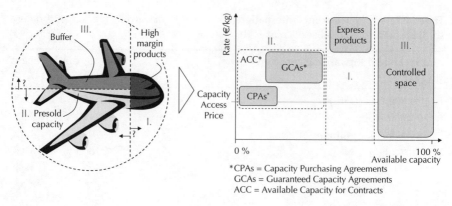

Fig. 2.5. Capacity allocation and reservation system

the development of joint processes as well as integrated IT and transaction systems. Beyond that, LCAG deliberately tries to create a "pull" for its services from end customers (via the forwarder of their choice) by its market positioning as a premium, high-quality carrier and product-innovation leader.

2.2.3 Revenue Management System

LCAG offers scheduled service on the basis of a six-month flight schedule. Once the schedule has been announced, the carrier markets the available capacity and tries to maximize revenues. Three different departments of LCAG's marketing division are involved in this process. The product management department governs the product offering. The revenue management department deals with demand forecast updating, capacity allocation, and operational pricing and generates input for network adjustment planning. It employs about 90 revenue analysts, flight analysts, revenue agents, and load planning agents. General pricing policy and strategy are the responsibility of the pricing department. Different from the traditional organizational structure of many airlines, this approach combines all pricing, product, and capacity decisions under one roof, namely the marketing division's.

2.2.3.1 Capacity Allocation

The core of LCAG's revenue management activities is the decision on how to split-up total available capacity (see Fig. 2.5, left hand side) into different allotments:

- *Express product* allotment. By nature, customers book these high margin products on short notice, i.e., a few days or even hours prior to departure. Since LCAG gives a guarantee of availability and on-time delivery for these products, it needs to forecast spot-market demand and reserve a respective portion of total capacity.
- *Available Capacity for Contracts* (ACC). This is done by signing long-term (one or two flight-schedule periods) contracts with forwarding companies and also determines how much capacity remains for general cargo in the spot market.

This translates into a capacity allocation chart as depicted in Fig. 2.5, right hand side. A large proportion of total capacity is pre-contracted in the form of *Guaranteed Capacity Agreements* (GCA). A GCA is a contract with a typical lifespan of six months between LCAG and a forwarding company. The contract reserves a certain amount of capacity on a particular route and day(s) of the week for the respective forwarder. The contract gives the forwarder the right to return unneeded capacity up to 72 hours prior to departure free of charge. If capacity is returned closer to departure, LCAG charges a cancellation fee of 25–100 % of the agreed rate, depending on the respective time of cancellation. However, due to the market power of some forwarding companies, this policy is not always strictly enforced.

Beyond that, LCAG offers *Capacity Purchasing Agreements* (CPA) to its forwarding customers. These contracts constitute the (non-returnable) obligation of the forwarder to purchase a fixed amount of capacity on a particular route and day(s) of the week. Forwarders, having traditionally been reluctant to firmly commit to any purchase obligation, mainly buy CPAs on high-demand routes, where capacity is in short supply and bottlenecks anticipated or where they are sure to have demand from shippers. Due to the fixed commitment, the price for CPAs to a particular destination is generally lower than for GCAs to the same destination. The typical duration of a CPA is 6 or 12 months with the contract being non-terminable over its lifespan.

The capacity share presold by these types of long-term contracts amounts on average to approximately one third of total capacity, but varies significantly from route to route within a range of 0 to ca. 70 %, depending on the expected demand for general cargo in the spot market at standard rates. Bookings at standard rates start 30 days prior to departure.

If due to unusually voluminous or heavy goods capacity limits of a flight are reached only in terms of volume *or* weight respectively, LCAG tries to find a shipper or forwarder with a complementary piece of cargo, i.e., of very high or very low density respectively. This way, LCAG can sometimes sell capacity twice and is therefore willing to grant favorable rates for such *ad hoc* deals.

In case of low demand, LCAG sells left-over capacity close to variable cost as *fill up*. On some routes, which Lufthansa serves with passenger aircraft, demand for air cargo does not correspond with demand for passenger transportation, so that lower-deck freight capacity is simply sold off.

Under this regime, overall capacity utilization[7] amounted to 65.6 % in the fiscal year 2003 (Lufthansa 2004), with a cargo load factor of 71.1 % on freighter aircraft and 57.2 % on passenger services, respectively (Lufthansa Cargo 2004).

2.2.3.2 Overbooking

Once total capacity has been allotted to the various market segments, LCAG's revenue management system plans for overbooking the capacity allotments, i.e., accepting in total more bookings for a flight than it can actually carry. Overbooking is motivated by no-shows that regularly occur. A no-show is cargo from a customer who has booked capacity but does not show up with the respective cargo. Since payment for shipments is generally not due before the time the airway bill is issued, LCAG does not receive remuneration for the blocked capacity and thus incurs an opportunity loss. However, overbooking includes the risk of having to offload cargo if eventually all or almost all cargo does show up for transportation. LCAG then has to compensate the respective customers, e.g., by paying contractual penalties, and, in the worst case, harms goodwill. The optimal overbooking level differs for each particular route and flight and is determined by using demand forecasts and estimates of the no-show probability from historic data.

2.2.3.3 Dynamic Pricing and B2B Trading Platforms

LCAG is working on the introduction of dynamic pricing (cf. Eye for Transport 2003). So far, the pricing department in coordination with the revenue management department determines the *Standard Rate Sheet* (SRS) for LCAG's particular products to each destination for one flight schedule period. Based on SRS rates, revenue analysts then set the rates for CPAs and GCAs. However, LCAG is considering switching to so-called bid prices that are continually adjusted to actual demand during the booking period of an individual flight (see also Sec. 3.3.3.3). The company has only recently made rates – both SRS and for long-term contracts – dependent not only on the destination of a flight but also on the day of the week it is scheduled, since LCAG has been experiencing different demand patterns for different days of the week. For example, demand on weekends is 10 times above weekday

---

[7] Capacity utilization = Tonne-kilometers taken (TKT) / Tonne-kilometers offered (TKO)

level for some destinations because forwarders consolidate shipments over the course of the week and ship in bulk on weekends.

The advent of dynamic pricing in the air cargo industry is believed to be accelerated by the emergence of electronic market platforms. In the air cargo industry, the trading platform *Global Freight Exchange* (GF-X) represent the largest business-to-business (B2B) marketplace. Founded in 1998, GF-X is a neutrally held wholesale platform that provides automated bookings and transactional functionality between air cargo carriers and forwarders (GF-X 2004). The platform provides information on schedule, capacity, rates, and bookings. Since 2003, LCAG has been offering forwarders with a long-term contract the possibility to place allotment bookings via the GF-X platform (GF-X 2003) (see also Fig. 3.2 on p. 36 for a classification of B2B exchanges).

2.2.4 Rationale and Problems of Long-Term Contracts

An important objective for LCAG to close long-term contracts, i.e., GCAs and CPAs, is the implied shift of capacity utilization risk. In the absence of advance-sale agreements, the asset provider would bear the entire capacity utilization risk. In the case of CPAs, which are non-returnable, the capacity utilization risk theoretically shifts entirely to the counter-party when the contract is signed. In return for the risk shift, LCAG grants lower prices for contract capacity (cf. FAZ 2001).

However, LCAG has to cope with various problems and challenges with regard to pricing and enforcing these types of long-term contracts:

1. Forwarders' reluctance about any form of fixed commitment: Reasons for this may on the one hand be grounded in mentality, on the other hand in strong economic reasoning. Spot price dynamics are hard to foresee and the chance of low spot prices reduces the willingness to purchase capacity in advance. This trend is aggravated on routes where overcapacity exists. Holloway (2003, p. 573) states on this problem: "[E]xcess output of cargo space on many medium- and long-haul routes (primarily in the belly-holds of passenger widebodies) has contributed to the [. . .] situation where ad hoc late sales are often made at lower-than-contract rates."

2. Pricing of long-term contracts: Overall, pricing of CPAs and GCAs is not done analytically. Especially for GCAs, LCAG finds it hard to value the flexibility inherent to the contract. The company feels that forwarders' freedom to cancel contracted capacity up to 72 hours prior to departure free of charge is currently not sufficiently reflected by the rates the airline charges for GCA freight. If a cancellation is received three days before departure, LCAG cannot be sure to find other buyers to make up for the foregone revenue. Furthermore, forwarders mostly ask for

CPAs on high-demand routes, while LCAG tries to limit the share of contracted capacity on such routes because capacity is likely to be sold at higher rates in the spot market.

3. Contract enforcement hindered by market power of intermediaries: Since payment for CPA capacity is made ex-post (i.e., after use of capacity), LCAG carries a substantial risk of forwarders defaulting on the contract and refusing payment for unused capacity. The same problem arises with the ex-post cancellation fee in the case of GCAs. Because of haggling and bargaining behavior of freight forwarders, LCAG find these fees oftentimes hard to collect.

Thus, the effective amount of risk carried by forwarders seems to be much less than the capacity price discount for CPAs suggests. In the case of GCAs, the intended risk shift is not achieved effectively either. The terms of contract, which allow customers to return agreed-upon capacity up to 72 hours prior to departure free of charge, still put LCAG in a position of carrying a substantial amount of capacity utilization risk.

The company has expressed considerations of introducing a split-tariff structure composed of an ex-ante reservation fee on signing the contract and an execution fee if the capacity is actually used, because an ex-ante collected reservation fee eliminates the risk of the counter party defaulting from the contract. Furthermore, since payment is partly received in advance, the capacity provider can use the funds in financing capacity provision (though this aspect will be ignored in Chap. 4). Such a pricing scheme is in the following referred to as a capacity-option contract. A forwarding company that signs such an option contract would acquire the right to use the agreed-upon capacity but could also let the option expire unused. The airline then would still make revenue through the reservation fee. LCAG expects capacity-option contracts to be especially advantageous in markets that exhibit high degrees of uncertainty in terms of demand and price volatility.

2.3 Research Questions and Methods

The problems with long-term capacity contracts are not unique to LCAG. The contract anomalies the company is experiencing have been commonly observed in the freight industry[8] and documented, e.g., by Pompeo and Sapountzis (2002, p. 92) who state:

"Contract practices in freight make future revenues extraordinarily unpredictable. Customers can reserve space on terms that in effect

[8] Potential applications in other industries are discussed in Sec. 8.2.

give them a free call option, for example. If the spot price for space falls below the forward price agreed upon between carrier and customer, the customer can simply rebook space on the spot market, without paying any penalty to the original provider. Even if the contract has a minimum-volume clause, such provisions are rarely enforced."

Under the standard industry practice of long-term firm-commitment (CPAs) and limited-flexibility contracts (GCAs) and with the presence of an alternative spot market, forwarders have an incentive to default from the first type of contract, while the inherent flexibility of the latter type of contract does not seem to be sufficiently reflected by contract prices, especially since carriers often cannot enforce the terms of the contract. Hence the risk-sharing objective of long-term contracts is not reached effectively. Instead, short-term cancellations result in high operational costs and opportunity losses for the airfreight carrier.

Starting from this status quo, the research question to be answered in the following is if capacity-option contracts as compared to fixed-commitment contracts represent a way to price flexibility adequately, allow for effective risk sharing, and thus lead to a more efficient market outcome.

To answer this question, the research question is broken down into the following three components:

1. How can capacity in a capacity-option contract be priced, i.e., what is the optimal pricing policy of the seller?
2. What is the optimal capacity reservation and execution policy of a buyer who is offered capacity options?
3. Under what environmental conditions will market participants prefer capacity-option contracts over fixed-commitment contracts, i.e., what drives the value of capacity-option contracts from the perspectives of buyer and seller?

To this end, the interaction between the asset provider and the intermediary in the contract market and their respective interaction with the spot market will be modeled analytically (Chap. 4). Using a multi-variate optimization approach, the optimal tariff for capacity reservations and thus the optimal pricing policy of the asset provider will be derived as well as the optimal reservation and execution policy of the intermediary. The model determines the optimal choice from the choice of either a fixed-commitment or a capacity-option contract. The optimal policy of the asset provider entails the allocation of capacity to contract and spot market. The analytical model will then be subject to a comparative static analysis in order to determine the influence of environmental conditions on the optimal policies of the market participants (Chap. 5).

In order to focus the analysis on the way of capacity reservation in the contract market, the setup of the model will necessarily have to abstract from some of the issues introduced in Sec. 2.2. These include:

- Multi-dimensionality of cargo: it will be assumed that cargo and capacity can be described by one single dimension (volume or weight) and is arbitrarily divisible.
- Segmentation of the spot market into an express-product and general-cargo segment: it will be assumed that capacity in the spot market is a homogenous good that is sold at an exogenously given market price.
- Uncertainty of capacity: total capacity will be assumed fixed and given.
- Network effects: interrelations between different routes and flights will be excluded.
- Non-enforcement of firm-commitment contracts: it will be assumed that the asset provider receives payment from the intermediary as agreed-upon in the capacity contract, be it a fixed-commitment or capacity-option contract.

3

Literature Review on Supply Contracting and Revenue Management

Having outlined in the previous chapter the modeling of capacity agreements in airfreight transportation as the scope of the thesis at hand, this chapter surveys the strands of literature related to this topic. This includes first of all the literature on *advance sale of capacity* that gives insight into the economic reasoning, implications, and modeling of capacity reservations. Secondly, the body of literature on *supply contracts* is of interest because it deals extensively with contract design and pricing in manufacturer–reseller settings. Thirdly, shedding light on the seller's perspective on capacity reservations and embedding these into the greater context of capacity management, select contributions to the existing literature on *revenue management* in general and on air cargo revenue management in particular are presented.

3.1 Advance Sale of Capacity

As described in Chap. 2, advance sale of capacity is standard practice in the air cargo industry. Air cargo carriers sell portions of a flight's total capacity en bloc to forwarding companies. In the following, it is discussed what motivates a capacity provider to sell capacity in advance, which implications result from doing so, and which modeling approaches on this topic exist in the literature.

3.1.1 Economic Reasoning and Implications

From both the seller's and buyer's perspective, advance sale of capacity can have multiple benefits. For the buyer, advance sale of capacity may, if it establishes a long-term supplier relationship, facilitate communication and information exchange, lower transaction cost due to less frequent rebidding for contracts, imply economies of scale in production, and consequently result in lower purchasing prices (cf. Cohen and Agrawal 1999). Furthermore, a long-term contract may represent a hedge against product price uncertainty if a fixed price is specified by the contract.

For the case of risk averse buyers, Png (1989) argues that reservations act as an insurance both against uncertainty of capacity availability and against uncertainty of the buyer's valuation of a product or service. The buyer's trade-off between lower procurement cost from using long-term orders and more accurate information on actual demand and valuation if using short-term orders is also made explicit by Burnetas and Gilbert (2001) for short-life-cycle products and modeled for information goods such as movie distribution rights or newspapers by Gundepudi et al. (2001).

Serel et al. (2001, p. 635) summarize that by using long-term contracts "[b]uyers seek to lower their purchasing costs, and have products delivered without interruption. [. . .] [S]uppliers are less pressured to find new customers and can afford to charge a price lower than the prevailing spot market price."

From the seller's perspective, Weatherford and Pfeifer (1994) point out that advance bookings provide information on total demand. Furthermore, sellers may use early-discount pricing to stimulate additional price-sensitive demand (this aspect is discussed in greater detail in the section on revenue management, Sec. 3.3). In this context, Chen (2001) develops an early-discount pricing strategy for a monopolist selling to heterogenous customers who self-select the price to pay by choosing the future shipping date of the product.

Shugan and Xie (2000) stress that applying advance sales to services implies separating purchase and consumption, which may cause customers to purchase in different quantities because at the time of purchase they act under uncertainty with regard to their valuation of the service at the time of consumption. The authors show that the seller can exploit this fact and improve his profits by using advance pricing if the buyer's valuation is sufficiently large compared to the service provider's variable cost. As Shugan and Xie point out, "[f]uture research might suggest how service providers could participate by selling options to buy future services at guaranteed prices" (ibid., p. 229).

Quan (2002) draws an analogy of capacity reservations to financial options (see also Sec. 3.2.3.2), arguing in the context of hotel room reservations that a hotel incurs a measurable (opportunity) cost by accepting a reservation while the reservation has a measurable value to the guest who can lock-in one room rate and search for a lower one. Quan finds this in contrast to the common practice of not charging guests for making or breaking reservations and attempts to value room reservations as European call options written on the price of a hotel room by applying the option-pricing model from Black and Scholes (1973).

3.1.2 Selected Modeling Approaches

In the following, selected modeling approaches of advance capacity sale are presented that provide insights into the sources of and influencing factors on the benefit of capacity reservations. Though in this regard similar to the results derived in this thesis, the following contributions regularly consider fixed-commitment (forward) contracts only.

From a seller's perspective, Weatherford and Pfeifer (1994) determine within the setting of the classic single-period inventory (Newsboy) problem (see Sec. 3.2.1.1) when a manufacturer should engage in advance bookings of orders by quantifying the benefits from demand stimulation and using advance bookings as a leading indicator of total demand.

For a firm that purchases supplies from an external manufacturer at deterministic prices and thus from a buyer's perspective, Serel et al. (2001) investigate the optimal action of a buyer who has a capacity-reservation contract in place with a preferred supplier and in addition an alternative spot market supplier at hand. The authors find that the amount of reserved capacity declines if alternative sourcing is available and that it decreases in demand uncertainty.

For a monopolistic service firm, Lee and Ng (2001) study the optimality of advance sale of capacity and show that the firm can earn higher profits and increase capacity utilization even though it conducts advance sales at a discounted price. Furthermore, the authors point out that the size of the advance sale allotment and optimal pricing depend on the expected price sensitivity at the time of consumption. If customers are less price sensitive at the time of consumption, it is optimal to allocate less capacity for sale in advance.

Gallego et al. (2003) model explicitly the impact of competition on advance sale of capacity by considering the market entry of a second capacity provider who sells at lower prices than the incumbent in both spot and forward market. The authors show that even in the absence of market segmentation the capacity providers engage in the discounted forward market if the ratio of total demand to total capacity is low, i.e., capacity is ample. If demand is high and capacity scarce, both capacity providers do not offer capacity in the contract market; for medium levels of capacity, mixed strategies apply.

Besides forward contracts, other types of contracts have been discussed in the literature for use of advance sale of capacity. These – as well as further insights on how to contract – are discussed in Sec. 3.2. The seller's perspective on capacity reservation and the integration thereof in a revenue management system is discussed in Sec. 3.3.

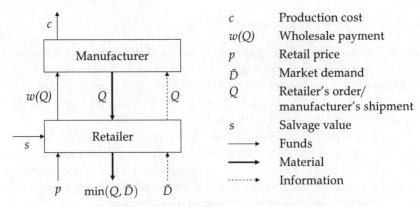

c	Production cost
$w(Q)$	Wholesale payment
p	Retail price
$\tilde{D}$	Market demand
Q	Retailer's order/ manufacturer's shipment
s	Salvage value
$\longrightarrow$	Funds
$\longrightarrow$	Material
------→	Information

Source: adapted from Tsay et al. (1999, p. 304)

Fig. 3.1. Structure of a one-period basic supply chain model

3.2 Supply Contracts

The supply chain management (SCM) literature treats environments in which multiple decision makers interact with each other for ordering and delivering a product. A typical example is the relationship between a manufacturer and a retailer. Fig. 3.1 shows the basic setting of such a supply chain, which is however not limited to a manufacturer–reseller relationship, but may represent any combination of an upstream and downstream party, e.g., a supplier and a manufacturer and their order and delivery relationship for a component or raw material.

To render the analysis of supply chains tractable, the SCM literature commonly reverts to the following simplifying assumptions (cf. Tsay et al. 1999, p. 303): In a one-period supply chain as depicted in Fig. 3.1, a retailer sells a product to the market at retail price p and receives a salvage value s for any unsold unit at the end of the selling season, with $p > s$. In anticipation of market demand $\tilde{D}$, the retailer orders Q units of the product from the manufacturer, who charges the retailer the wholesale payment $w(Q)$ per delivery and produces (or acquires) the product at unit cost c. In reality, market demand $\tilde{D}$ is both price-sensitive and uncertain. Though some models do consider both these features, it is common in the operations research literature to assume the retail price as fixed and market demand as stochastic (indicated here by a tilde on the demand variable). In the economics and marketing literature, it is more common to assume a deterministic, downward-sloping demand function (ibid., p. 303). In the latter case, the primary decision variable of the retailer is the retail price p, in the former the order quantity Q. The structure and amount of the wholesale payment

$w(Q)$ depends on the type and terms of the contract between manufacturer and retailer.

The approach taken in the model introduced in Chap. 4 corresponds to the commonly used framework in the operations research literature and deals with the supply chain parties' decisions on order quantity and wholesale payment. The remainder of this section focuses on contributions of the SCM literature that provide insights into modeling these decisions. One of the standard building blocks for modeling order quantity decisions under stochastic demand is the Newsboy model, which represents the basis for most discrete-time stochastic inventory models (cf. Lee and Nahmias 1993, p. 26). Because the structure of the buyer's problem in Chap. 4 also corresponds to a Newsboy-like situation, the genesis and formulation of as well as extensions to the standard Newsboy model are introduced in the following subsection. Subsequent to this, purposes of supply contracts, different contract types, and their application for capacity reservation are discussed.

3.2.1 Order Quantity Decisions Under Stochastic Demand: The Newsboy Model

The Newsboy[1] model addresses order quantity decisions under stochastic demand. It is widely used in modeling advance sale of capacity (see, e.g., Weatherford and Pfeifer 1994; Burnetas and Gilbert 2001) and revenue management. A recent review on the Newsboy problem is provided by Khouja (1999) who presents various extensions to the model. Those of relevance to the model in Chap. 4 are discussed further down in this section.

According to Petruzzi and Dada (1999), the logic of the Newsboy model dates as far back as to Edgeworth (1888) who discusses the cash-flow problem of a bank as the balance between "a slight chance of a great loss against a good chance of a few shillings" (ibid., p. 121). By that Edgeworth described what was later to be coined the *critical fractile* that balances *underage* and *overage* cost. The first formulation of the model with respect to inventory management can be found in Arrow et al. (1951). The significance of this work is documented by Arrow (2002).

3.2.1.1 The Standard Model

The Newsboy model in its simplest form is a one-period one-product inventory model. The seller faces the problem of determining the optimal order quantity Q of a product that satisfies the following set of assumptions:

---

[1] Both the terms *Newsboy* and *Newsvendor* model can be found in the literature (cf. Nahmias 2004; Porteus 1990), with the first being the more classic, the latter the more politically correct expression. In this text, they are used interchangeably.

- The product is a perishable product, i.e., it can be used to satisfy demand during that period only.
- No procurement is possible over the selling season, i.e., all items have to be purchased at the beginning of a period.
- Demand for the product is uncertain. More specific, demand $\tilde{D}$ is a continuous nonnegative random variable with density function $f(\tilde{D})$, cumulative, invertible distribution function $F(\tilde{D})$, mean μ, and standard deviation σ.
- All prices and cost parameters are know with certainty.

The vendor purchases the product at the unit purchase price w from its supplier and sells it at the unit selling price p to its customers. Every unsold item has a salvage value of s. It is assumed that $p > w > s \geq 0$. The vendor thus incurs a cost for every unit that is unsold at the end of the period, namely $w - s$. This cost is called the *overage cost* c_o. If the vendor has ordered too few items, he incurs a shortage cost, in this case the opportunity cost of $p - w$ for every unit of unsatisfied demand. This cost is called the *underage cost* c_u.

This situation resembles the daily decision problem of a newsboy who has to decide in the morning on how many newspapers to buy, knowing that any copy unsold by the end of the day is worth only its waste-paper value. Thus the name of the model.

The model formulation presented here is based on Nahmias (2001, pp. 241 ff.), Porteus (1990, pp. 610 ff.), and Silver et al. (1998, pp. 387 f.). The latter show that the cost minimization approach that is usually chosen in standard textbooks is equivalent to the profit maximization approach presented in the following (and which is used later in Chapter 4).

The problem is analyzed by deriving the profit function of the newsvendor as an expression of $\tilde{D}$ and Q. Let P denote the Newsvendor's profit and $\mathcal{E}[P] \equiv \Pi$ (for notational convenience) the expected value thereof. The profit function is:

$$P = p \min(\tilde{D}, Q) + s(Q - \tilde{D})^+ - wQ \tag{3.1}$$

where $(Q - \tilde{D})^+ = \max(0, Q - \tilde{D})$. Using of the relationships $\min(\tilde{D}, Q) = \tilde{D} - (\tilde{D} - Q)^+$ and $Q = \tilde{D} - (\tilde{D} - Q)^+ + (Q - \tilde{D})^+$ (cf. Rudi and Pyke 2000, p. 172), rearranging terms, and substituting $w - s = c_o$ and $p - w = c_u$ gives:

$$P = (p - w)\tilde{D} - c_o(Q - \tilde{D})^+ - c_u(\tilde{D} - Q)^+ \tag{3.2}$$

The expected value of this expression is:

$$\Pi = (p - w)\mu - c_o \int_0^Q (Q - \tilde{D})f(\tilde{D})d\tilde{D} - c_u \int_Q^\infty (\tilde{D} - Q)f(\tilde{D})d\tilde{D} \tag{3.3}$$

The value of Q that maximizes expected profits, denoted as Q^*, is determined by setting the first derivative with respect to Q equal to 0 and solving for Q (first order condition).

$$\frac{\partial \Pi}{\partial Q} = -c_o \int_0^Q 1 f(\tilde{D}) d\tilde{D} - c_u \int_Q^\infty (-1) f(\tilde{D}) d\tilde{D} \tag{3.4a}$$

$$= -c_o F(Q) + c_u \left(1 - F(Q)\right) \tag{3.4b}$$

$$\overset{!}{=} 0$$

Rearranging terms gives

$$F(Q^*) = \frac{c_u}{c_o + c_u} \tag{3.5}$$

It follows directly from (3.4b) that $\partial^2 \Pi / \partial Q^2 = -(c_u + c_o) f(Q) < 0$ (second order condition). Thus, $\Pi(Q)$ is a concave function that has a unique maximum Q^*.

The right hand side of (3.5) is called the *critical fractile*. Since c_u and c_o are by the above assumptions positive numbers, the fractile is always between 0 and 1. It is the probability that the entire period demand can be satisfied if the newsvendor purchases Q^* units of the product at the begin of the period. It can thus be interpreted as the optimal probability of not stocking out (Porteus 1990, p. 611) and marks the newsvendor's *service level*.

3.2.1.2 Extensions to the Standard Model

In the model introduced in Chap. 4, the decision of an intermediary who reserves capacity with an asset provider will be shown to be a Newsboy-like situation, however, with some extensions to the above standard model. These include the nature of the product, the type of demand distribution, the concept of demand, the nature of the shortage cost, and the structure of the wholesale payment. Contributions to the literature that deal with these aspects are presented in the following.

With regard to the *nature of the product*, it is widely acknowledged that services, as considered in Chap. 4, show – in the context of a Newsboy-type model – the same characteristics as perishable products (Khouja 1999, p. 550). Indeed, services exhibit the feature of expiry at the end of the sales period in its purest form because the revenue from service capacity, e.g., a seat on a plane, that is unsold at the plane's departure, can never be recovered.

The Newsboy model has been considered in the literature for various *types of demand distributions*. One of the most commonly used is the normal distribution, which is likewise applied in the model in Chap. 4. For this type of distribution it is possible to find near-closed-form expressions for the expected profits of buyer and seller (cf. Lau 1997; Silver and Peterson 1985, p. 697), which render the comprehensive numerical analysis conducted in Chap. 5 feasible.

Concerning the *concept of demand*, the standard Newsboy model assumes exogenous demand. Demand can be made endogenous by assuming price-dependent (and, as before, random) demand, which allows to focus on the coordination of replenishment strategy and pricing policy. Among the first to analyze the Newsboy problem with price-dependent demand have been Whitin (1955) and Mills (1959) who provide explicit calculations for the special case of a uniform demand distribution. Lau and Lau (1988) show that, though the critical-fractile solution continues to hold in the case of price-dependant demand, analytical solutions for both the optimal stock level and optimal price cannot be found for others than the simplest price-demand relationships. The model introduced in Chap. 4 recurs to a formulation of the demand function (additive demand) from Petruzzi and Dada (1999). To solve the model for the optimal stock level and the optimal price, the authors propose a sequential procedure that first determines the optimal stock level as a function of the price, then substitutes the result back into the seller's objective function, and finally determines the optimal price by a numerical search over the objective function. It will turn out that this procedure is also applicable to the model presented in Chap. 4.

While, in the standard model, all cost parameters are known with certainty, this assumption might not always adequately reflect reality. Especially *shortage cost*, since these may include hard-to-measure components like loss-of-goodwill cost, may not be known with certainty. In the model in Chap. 4, the shortage cost will be the function of a stochastic variable. Uncertainty of shortage cost has been studied by Ishii and Konno (1998) who present a Newsboy model where shortage cost are represented by a fuzzy number. The authors find that the optimal order size increases as the ambiguity of shortage cost becomes larger.

Instead of using a linear wholesale price as in the standard Newsboy model, different *structures of the wholesale price* have been discussed in the literature. For example, Khouja (1996) discusses supplier quantity discounts and their influence on the buyer's optimal ordering policy. Rudi and Pyke (2000) formulate a Newsboy model with options where the retailer buys Q call options at unit cost r (reservation fee) from the manufacturer. Each call option gives the retailer the right to purchase one unit of the product from the manufacturer at execution fee x after observing demand.[2] By choosing r and x appropriately, the retailer can be induced to order a higher quantity as

---

[2] The nomenclature presented here slightly differs from Rudi and Pyke (2000) to avoid ambiguity with previously introduces variables.

compared to the standard model's pricing scheme with a single wholesale price while both manufacturer and retailer earn higher expected profits.

The model in Chap. 4 is based on the formulation by Rudi and Pyke (2000). The decisive technical difference to their model is the determination of the optimal tariff which is absent in Rudi and Pyke (2000) because the authors assume exogenous demand. Furthermore, going beyond Rudi and Pyke (2000), the model formulation in Chap. 4 includes multiple sources of uncertainty and an alternative market.

3.2.2 Purposes of Contracts

Tsay et al. (1999, p. 304 ff.) distinguish three purposes of supply chain contracts: system-wide performance improvement, risk sharing, and facilitation of long-term partnerships. The following discussion of these motives is aimed at helping to structure the subsequent analysis of various contract types.

In an ideal world, the quantity and pricing decisions in the supply chain as shown in Fig. 3.1 would be made by a single decision maker who has all information at his disposal. This situation is usually referred to as the *centralized supply chain* and the decision maker the *integrated firm*. In real-world supply chains, usually multiple decision makers interact who exert local control and optimization. However, locally optimal behavior can be inefficient from the perspective of the entire system (cf. Whang 1995). One such effect is *double marginalization* (cf. Spengler 1950) which occurs in a supply chain as depicted in Fig. 3.1 if a linear wholesale price greater than the cost of production is used (Tirole 1988, p. 174 f.). The two successive mark-ups above the production cost charged by manufacturer and retailer result in the retailer's order quantity and thus the sum of the two parties' expected profits being lower than in the case of the integrated firm.

An important objective of supply chain contracts is thus *system-wide performance improvement*. By structuring the contractual agreement between manufacturer and retailer accordingly, it is possible to coordinate the decisions of the supply chain partners and ideally attain the result of a centralized supply chain while maintaining a decentralized structure (for an extensive review with regard to the use of contracts to achieve channel coordination, see Cachon 2003). The performance-improvement objective is common to many of the contract types discussed in the next subsection.

Even if a contract or wholesale price structure achieves channel coordination, this does not necessarily mean that the proceeds from the contract are split among the supply chain partners in a way that each party is willing to accept. This aspect is closely linked to how the parties share the risks arising from the uncertainty[3] associated with the supply chain, e.g., with regard to market demand, retail price, exchange rates, product quality, or lead time (cf. Tsay et al. 1999, p. 305). *Risk sharing* is thus a further motive that is pursued by closing supply chain contracts and discussed here.

Finally, a part of the literature focuses on the contracting motive of *facilitating long-term partnerships*. This includes especially the microeconomic perspective on contracts from institutional economics, based on the theory of incomplete contracts as initiated by Coase (1937). This has later become a cornerstone of principal agent (cf. Jensen and Meckling 1976) and transaction cost theory (cf. Williamson 1979). By entering persistent business partnerships, buyer and seller can reduce transaction costs since costly searches and renegotiations are reduced. Furthermore, e.g, by explicitly specifying penalties for non-cooperative behavior or implicitly because agents rationally incorporate the long-term effects of their present behavior into their decision making, principal-agent conflicts can be alleviated. This perspective on contracts is not the focus of the thesis at hand and will thus not be further pursued in this review. The same applies to the legal perspective on contracts that is, for instance, dealt with in Triantis (2000).

3.2.3 Contract Types

Two kinds of supply contracts can be distinguished in the light of the long-term capacity purchasing and reservation contracts, respectively, introduced in Chap. 2 and modeled in Chap. 4: non-flexible and flexible contracts. Non-flexible contracts designate in the following contracts under which the buyer needs to make a one-time decision about the quantity delivered by the seller at a predetermined price before uncertainty is resolved. Flexible contracts in contrast allow for adjustments of price and/or quantity after the initial order has been placed and some or all uncertainty has been resolved. With the essential feature of the capacity agreements dealt with here being flexibility, the focus of this subsection is on flexible contracts. Before different types of these are investigated in greater detail, first a common split-tariff pricing scheme including also non-flexible contracts – the two-part tariff – is discussed in order to draw a dividing line between this tariff scheme and the price components in an option contract.

---

[3] Knight (1921) distinguishes between *uncertainty* and *risk*. He associates risk with a known probability distribution over unknown outcomes and uncertainty with situations where even the probability distribution is unknown. If one strictly followed this terminology, the analysis conducted in subsequent chapters and in most of the works cited in this chapter would deal exclusively with decision making under *risk*.

3.2.3.1 Contracts with a Two-Part Tariff

The tariff structure of the option contract modeled in Chap. 4 is composed of two parts, the reservation fee r and the execution fee x, and can as such be called a two-part tariff. This term, however, has in the literature traditionally been used for a wholesale-payment or retail-price structure composed of a fixed (i.e., volume-independent) payment F plus some linear unit price w, such that (for the example of a wholesale payment as in Fig. 3.1) $w(Q) = F + Qw$.[4] Familiar examples for the application of such tariff structures include telecommunication services (cf. Blonski 2002) and utilities such as gas, water, and electricity. Here, the pricing by providers usually includes a monthly or yearly basis charge plus some consumption-dependent unit charge.

Especially with regard to retail price setting, two-part tariffs have been discussed extensively in the literature, e.g., by Oi (1971) who investigates profit maximization in a monopoly and, under more general conditions, by Skiera (1999, p. 83–104). Spremann and Klinkhammer (1985) identify several circumstances under which two-part tariffs can be observed. As a required condition, the authors put forward that it must not be possible for the buyers to resell the product to other consumers in order to prevent arbitrage through the formation of buying communities (ibid., p. 792). Comparing a two-part tariff with option contracts as proposed in Chap. 4, two decisive differences apply: Firstly, in the option contract, both payments (at reservation and at execution) are volume-dependent, in the two-part tariff only one. Secondly, when considering a setting as depicted in Fig. 3.1, the application of a two-part tariff as wholesale payment would still imply the buyer to make the decision about the quantity delivered by the seller before demand is realized; in the case of the option contract, the buyer makes this decision with knowledge about actual demand.

It has been discussed, primarily in the marketing literature, if two-part tariffs can be a means to coordinate the supply chain. Moorthy (1987) demonstrates in a manufacturer–retailer setting how a two-part tariff can achieve channel coordination while Jeuland and Shugan (1983, 1988) argue that this approach does not hold if each channel member is to keep control over his respective decision variables. In a recent working paper, Ho and Zhang (2004) report on the results of a laboratory experiment on the performance of two-part tariffs and conclude that even in well-controlled market environments a two-part tariff structure fails to solve the double-marginalization problem.

---

[4] To avoid ambiguity, the term *two-part tariff* will in accordance with its use in the literature be reserved in this text for the combination of a fixed and linear price component. The combination of a reservation and execution fee will later sometimes be referred to by the more general term *split tariff*.

3.2.3.2 Flexible Supply Contracts

Flexible supply contracts allow for a modification of quantities and/or prices contingent on the resolution of uncertainty. Flexible supply contracts thus respresent a form of contingent contracts. Bazerman and Gillespie (1999) illustrate by formulating the terms of a contract explicitly as dependent on the future outcome of an ex-ante uncertain event that the contract parties can build trust on and protect themselves from one party earning ex-post windfall profits at the expense of the other. In addition, Bazerman and Gillespie list requirements for the effectiveness of contingent contracts. Of these, the most important in the context of the capacity agreements discussed in this thesis is the enforceability of the contract in order to prevent the parties from defaulting ex post. In line with this observation, the enforceability of Guaranteed Capacity Agreements at LCAG has been identified as a major obstacle in Chap. 2.

The desire for flexibility in supply contracts arises from the uncertainties associated primarily with, but not limited to, demand and prices. Further uncertainties include quality, capacity, exchange rates, cost, and lead time. In the following, the focus is on the management of demand and price uncertainty. Flexibility can both serve as a hedge against unfavorable outcomes and be used to maintain the chances associated with favorable outcomes. For example, a flexibility clause may be beneficial for the buyer in the case of low demand when he prefers a lower delivery quantity rather than his initial order or in the case of low spot price when he prefers to purchase on the spot market rather than from his contracted supplier. Flexible contracts have been shown to be an effective means to achieve the contract purposes of system-wide performance improvement and risk sharing (see Sec. 3.2.2).

Quantity Flexibility Contracts

In the presence of demand, cost, and capacity uncertainty, Spulber (1992) shows that a *non-linear tariff* can lead to a Pareto efficient outcome, i.e., a system-wide performance improvement. This is achieved by introducing a base-load demand chosen ex ante by the buyer that determines the ex-post payment level and allocation rule.

In a more standardized fashion, *quantity flexibility* (QF) contracts specify the terms under which the buyer's actually received quantity may deviate from the original order quantity Q. The clauses can include limits on the range of allowable changes, pricing rules, or both (Tsay et al. 1999). Examples for QF contracts include backup agreements as modeled by Eppen and Iyer (1997) for fashion apparel goods. Prior to the selling season, the buyer places his original order with the manufacturer, but gets delivered only a fraction of this order. After having observed the first two weeks of sales, the buyer decides if he wants to receive the entire remainder of the original order or only a part of it. In the latter case, the manufacturer charges the buyer a penalty cost for each unit of the original order that the buyer does not take delivery of. An in-depth analysis of QF contracts is provided by Tsay (1999) who focuses on the incentive of buyer and seller to participate in a QF contract which is constituted by a trade-off of price and flexibility. He points out that the buyer is willing to commit to a certain order quantity in exchange for a lower price while the seller benefits from more predictable sales and therefore is willing to grant a discount.

While QF contracts may offer *full* refund for a *partial* fraction of the original order, Lariviere (1999) shows the close link to another kind of flexible contracts, namely *buy-back contracts* with return policies, which may offer *partial* refunds for returning, in extreme cases, the *full* original order.

Buy-Back Contracts, Return Policies, and Revenue Sharing

A return policy allows the retailer to return unsold items to the manufacturer and collect a repurchase price per unit returned, which is lower than the wholesale price. The manufacturer thus commits to buying back unsold items. From a managerial perspective, Padmanabhan and Png (1995) discuss under what conditions buy-back contracts are most likely to be observed. The first to analytically treat buy-back contracts was Pasternack (1985) (in the marketing literature) who shows that neither a policy that allows for unlimited returns at full credit nor a no-refunds policy is optimal. Instead, channel coordination can be achieved by a return policy allowing for unlimited returns at partial credit with a combination of wholesale price and repurchase price chosen from a range of optimal value pairs, each of which exhibits a different profit split between retailer and manufacturer.

In this context, Lau and Lau (1999) address the problem of the profit split and conclude that the manufacturer does not necessarily choose a combination of wholesale and repurchase price from Pasternack's optimal set because these values can be unfavorable to the manufacturer. However, even in these cases, Lau and Lau (1999) show that a buy-back contract achieves a Pareto improvement. A generalization of the model by Pasternack (1985) can be found in Emmons and Gilbert (1998) who incorporate price-dependent demand and can thus in addition determine the optimal retail price.

For the standard Newsvendor case with exogenous demand, Cachon and Lariviere (2002) show that results equivalent to a buy-back contract can be established if the manufacturer and the retailer agree upon *revenue sharing* instead of a return policy. A revenue sharing contract is specified by two parameters: the per-unit wholesale price w and the retailer's share of revenue γ. The manufacturer receives $1 - \gamma$ of the revenue that is assumed to be made only by the retailer. As in the buy-back contract, any arbitrary profit split can be achieved by choosing w and γ. For a revenue-sharing contract to work, the manufacturer must be able to effectively monitor the retailer's sales and sales effort and may therefore bring with it a substantial administrative burden.

For the original buy-back contract formulation from Pasternack (1985) with exogenous demand, Rudi and Pyke (2000, see also Sec. 3.2.1.2) show that equivalent results can be obtained by re-interpreting the partial-refund contract as a set of European call options with reservation fee r and execution fee x. Equivalence with Pasternack's buy-back contract can be established with the wholesale price equal to $r + x$ and the repurchase price equal to x. Accordingly, combinations of r and x can be found that coordinate the channel when the retail price is fixed and demand exogenous. Apart from this analogy, however, buy-back contracts differ decisively from option contracts with regard to the production of the traded goods: With a buy-back contract, production and delivery has typically already occurred when the original purchase quantity is adapted; in an option contract, delivery and perhaps even production take place only after option execution.

Flexible Supply Contracts with Options

Before continuing the discussion of flexible contracts with a review of supply contracts with options, first a brief account on the background of options in the finance and operations management literature is provided. This includes financial options and real options.

Background from Options Theory

Options theory has initially developed primarily with regard to the valuation of *financial options*. A financial option is the right, but not the obligation to purchase (call option) or to sell (put option) the underlying financial asset, e.g., corporate stock, at (European option) or up to (American option) a certain date at a predetermined price (the *strike price* or *exercise price*) (cf. Hull 2003, p. 6). The by now standard method for determining the value of financial options and thus the option price has been proposed by Black and Scholes (1973). They show that the price of a financial option depends on the current stock price of the underlying, the strike price, the time to maturity, the risk-free rate of return, and the volatility of the stock price.

Options involving real (as opposed to financial) assets, such as land, a manufacturing plant, or machinery, have been coined *real options* (cf. Kester 1984; Brealey and Myers 2000, p. 619). Real options arise from recognizing the value of the flexibility of altering decisions over time. Real options are classified primarily by the type of flexibility that they offer and include, for instance (cf. Trigeorgis 1995; Copeland and Antikarov 2001), the option to defer (e.g., delay the start of a project), the option to abandon (e.g., stop a project and realize its resale value), the option to switch (e.g., change the operating mode of a plant), and the option to improve (e.g., realign development goals in a R&D project to changing consumer preferences) (cf. Huchzermeier and Loch 2001). Real-option valuation techniques are used in capital budgeting and the management of R&D projects (cf. Dixit and Pindyck 1994; Luenberger 1998). Because of the inherent complexity of real-option decision problems, the most widely used valuation techniques include lattice approaches (cf. Amram and Kulatilaka 1999, p. 113 ff.), e.g., binomial option pricing (cf. Cox et al. 1979).

Options used in supply contracts, in the following referred to as *supply options*, likewise constitute the right, but not the obligation to receive (call option) or deliver (put option) a good, product, or service by a certain date and predetermined price. Though financial, real, and supply options thus share some common ground, underlying market assumptions and valuation techniques differ significantly. For supply options, the typical notion in the operations management literature is – consistent with other types of contracts discussed up to here – that prices and contract terms arise through a negotiation framework between most commonly a single seller and a single buyer rather than through a complete-market, equilibrium approach as in the finance literature (cf. Kleindorfer and Wu 2003, p. 1607).

Options in Supply Contracts

For a supplier selling raw material or components to a manufacturer, Cheng et al. (2002) model a flexible supply contract in which the buyer is required to firmly commit to one part of the order quantity and purchases options on the remainder. By considering call and put options, the authors point out the analogy to buy-back and return policies. Assuming the base price and quantity of the firmly-committed order part as given, Cheng et al. derive the seller's optimal reservation and execution fee and the buyer's optimal number of options to purchase and show that channel coordination can be achieved when the contracting parties negotiate over the profit-sharing mechanism.

Likewise for a supplier–buyer setting, Barnes-Schuster et al. (2002) investigate *supply contracts with options* in a two-period model. The supplier can operate in two production modes: a cheaper production mode with a longer lead time and an expensive mode with a short lead time. The authors show that both backup agreements and QF contracts (see above) are special cases of their general model and derive sufficient conditions on the cost parameters that allow for channel coordination. They show that prices that achieve channel coordination may violate the individual rationality constraint for the seller[5] and demonstrate numerically the system-wide performance im-

---

[5] This corresponds to the above cited finding by Lau and Lau (1999) for the coordinating returns policy (see Sec. 3.2.3.2).

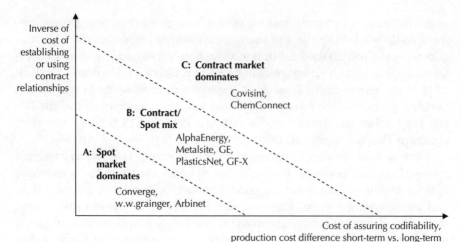

Source: adapted from Kleindorfer and Wu (2003, p. 1599); positioning of GF-X
added

Fig. 3.2. Classification of B2B exchanges and contracting

provement achieved by options. An implementation of the model by Barnes-Schuster et al. (2002) via stochastic programming is provided by van Delft and Vial (2004) that allows for numerical analysis of various contractual parameters.

Kleindorfer and Wu (2003) survey the theory and practice in the use of options in business-to-business (B2B) exchanges and investigate how options have been used as a means to integrate long- and short-term contracting. Especially the emergence of electronic market platforms such as B2B exchanges that enable last-minute spot market purchases has recently fueled the trend towards combining different procurement forms. Kleindorfer and Wu (2003) also classify existing B2B exchanges with regard to the predominant contract form and establish a classification framework based on the cost of assuring codifiability (i.e., the ability to electronically specify product, delivery, and settlement requirements in a verifiable manner), the difference in production cost between spot- and contract-based sales on the one side and the cost of establishing or using contract relationships on the other side determine whether spot market transactions, contract market transactions, or a mixture of both are the predominant transactions in an electronic B2B exchange (see Fig. 3.2).

For supply option contracts, electronic marketplaces may also serve as a platform for a secondary market (cf. Lee and Whang 2002), in which an option buyer can sell options which he foresees not to exercise to other buyers.

Likewise along the lines of mixing short- and long-term contracting strategies, Martinez-de-Albéniz and Simchi-Levi (2002) show how a manufacturer can increase expected profit and reduce his financial risk by using a portfolio approach for procurement contracts consisting of a combination of fixed-commitment contracts, option contracts, and spot market purchases. Martinez-de-Albéniz and Simchi-Levi (2003) investigate competition in the supply option market by developing a formulation for option contracts for physical delivery of electricity with one buyer and many suppliers.

In the field of capital-intensive production of non-storable goods and services, which – given the high share and absolute height of fixed cost – includes air cargo transportation services, Wu et al. (2002) develop bidding strategies in the presence of a long-term contract market and a short-term spot market, incorporating only market risk in the form of spot price uncertainty and assuming demand and cost as deterministic. The authors show that the seller's optimal strategy entails setting the execution cost as low as possible, i.e., to reveal marginal production cost, while extracting his margin from the reservation fee.

Spinler (2003) extends the model by Wu et al. (2002) to a general valuation framework by incorporating state-contingent demand and cost and a willingness-to-pay function likewise depending on the state of the world. Spinler derives analytical expressions for the buyer's optimal reservation quantity and the seller's optimal tariff and shows that the combination of option contracts and a spot market is Pareto improving. If the option buyers are allowed to resell options prior to execution in a secondary market, the buyers can engage in capacity-slot trading. In Spinler et al. (2003), the integration of option contracts for physical delivery and complementary financial derivative instruments into a company's risk management approach is discussed.

For delivery of electricity or natural gas in energy markets, Jaillet et al. (2004) present a numerical valuation scheme for swing options. Swing option contracts are a kind of variable volume contract that permits the option holder to repeatedly exercise the right to receive greater or smaller amounts of energy, subject to daily as well as periodic constraints. The authors show that hedging via swing options in the case of spiky, but back to normal reverting demand is more efficient than hedging via a series of European or a set of American options.

3.3 Revenue Management

This section aims at introducing the field of revenue management by focusing on those tools and concepts that are relevant to the subject of this thesis. Recent overviews of the extensive body of literature on revenue management (RM) in theory and practice are provided by McGill and van Ryzin (1999), Tscheulin and Lindenmeier (2003) and Talluri and van Ryzin (2004); Weatherford and Bodily (1992) provide a comprehensive taxonomy and research overview.

The field of revenue management subsumes approaches of dynamic and simultaneous price and capacity management (Tscheulin and Lindenmeier 2003, p. 630) and originates from the passenger airline industry. In the literature, especially in early papers on the subject, revenue management is also referred to as *yield management*. Some authors, though, have argued that the term *yield* (meaning, as a technical term in the airline industry, the actually obtained price) does not reflect the aspect of controlling price *and* quantity simultaneously (Weatherford and Bodily 1992, p. 833) and therefore prefer the term *revenue* (i.e., price *and* quantity) management. Today, the two terms are considered synonymous (cf. McGill and van Ryzin 1999, p. 251). Weatherford and Bodily (1992) have proposed the term *perishable asset revenue management* (PARM) for the optimal revenue management of perishable assets though price segmentation, which is thus explicitly not restricted to the airline industry. In this thesis, the term revenue management is used, without a specific industry context in mind.

3.3.1 Background

Revenue management techniques were first developed and applied in the passenger airline industry, with American Airlines being one of the pioneers in the 1960s (cf. Smith et al. 1992). More and more airlines started offering restricted fare products by the early 1970s, e.g., early-bird bookings that enabled airlines to gain revenue from seats that otherwise – at full fare – would have remained unsold (cf. McGill and van Ryzin 1999, p. 233 f.). From this practice, the need arose to control the mix of fares sold for a particular flight (cf. Belobaba 1987, p. 63) and to protect seats for late booking, full fare passengers. The class of optimization models concerned with this problem is titled *seat-inventory control*. Before and independent of this development, quantitative research had almost exclusively focused on the issue of *overbooking control*, motivated by airlines that wanted to predict and anticipate cancellations of bookings and no-shows of passengers (for greater detail, see Sec. 3.3.3.4). The deregulation of the U.S. passenger airline industry with regard to prices and schedules in 1979 further fueled research and application of revenue management tools.

The understanding of airline revenue management prevailing today comprises the entirety of these and other practices that are geared towards "controlling the availability and/or pricing of travel seats in different booking classes with the goal of maximizing expected revenues or profits" (McGill and van Ryzin 1999, p. 250). More generally and holistically, Harris and Pinder (1995, p. 299) state that "the basis of revenue management is an order acceptance and refusal process that integrates the marketing, financial and operations functions to maximize revenue from pre-existing capacity".

3.3.2 Criteria for Successful Applications

Harris and Pinder (1995), Kimes (1989), and Weatherford and Bodily (1992) have identified several common characteristics of businesses and situations in which revenue management is effectively applied:

1. *Perishability.* The product or service is non-storable and/or dated. If, in the context of the passenger airline industry, a seat is empty when the aircraft takes off, the revenue potential associated with this seat is lost for good.
2. *Fixed capacity.* At least in the short run, capacity cannot be adjusted to market demand. For example, when a flight is fully booked, an aircraft cannot be enlarged or easily be replaced by a larger one. However, it might exist some limited flexibility, e.g., of shifting passengers to a later flight.
3. *Low marginal cost, high fixed cost.* Marginal sales cost of selling one extra unit of inventory must be low, while capacity change costs are high; e.g., taking one additional passenger aboard is inexpensive, while providing additional capacity is very expensive since it is associated with high fixed cost.
4. *Segmentable demand structures.* The very idea of revenue management is to segment customers by their respective marginal willingness-to-pay. The product or service is then priced and marketed to each segment separately. In order to avoid dilution of the segments, i.e., to hinder price-insensitive customers from obtaining the product at the lower price geared to price-sensitive customers, the market segments need to be separable by geographic or demographic factors, by the time of purchase, by imposing restrictions on the product (e.g., Saturday night stay for discount airfares), or similar mechanisms.
5. *Advance sales/bookings.* Because booking patterns regularly correlate with price differentiation, i.e., customers that book late are usually more price-insensitive than customers to reserve capacity early in advance, the capacity provider needs to track capacity bookings and has to trade off the advantage that an advance sale assures capacity utilization with the disadvantage that early commitment blocks capacity that potentially can be sold later at a higher price.
6. *Stochastic and fluctuating demand.* If demand were deterministic and flat, the capacity provider could adjust capacity to demand. In the presence

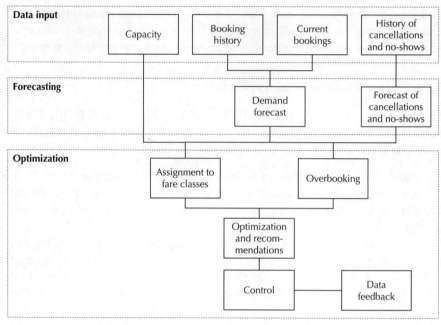

Source: adapted from Simon (1992, p. 585)

Fig. 3.3. Generic structure of an integrated revenue management system

of demand uncertainty, revenue management enables companies to exploit revenue opportunities in times of high demand (by raising prices) and to increase capacity utilization in times of low demand (by lowering prices).

7. *Historic sales data and demand forecasting.* To realize the full benefits of revenue management, companies need historic data and demand forecasts on which to build their decisions about allotment sizes, prices, overbooking levels, etc.

3.3.3 Components of Revenue Management Systems

Revenue management systems serve the simultaneous control of price and capacity. In an RM process, however, the control problem is usually divided in multiple subproblems and solved sequentially (cf. Stuhlmann 2000, p. 243, and Tscheulin and Lindenmeier 2003, p. 631). Fig. 3.3 gives an overview of the generic structure of an integrated revenue management system. The process can be divided into three parts: data input, forecasting, and optimization. The optimization part comprises price and capacity control, the latter being achieved by the application of models for inventory control and overbooking. The contributions to the RM literature introduced in the following focus on forecasting, pricing, inventory control, and over-booking.

3.3.3.1 Forecasting

Demand forecasts can be distinguished by their time horizon. Revenue management usually focuses on short-term forecasts, i.e., with regard to the time horizon for which capacity is fixed. For an airline, this is usually the duration of a flight schedule (6–12 months). In the mid- and long-term, capacity can be adjusted, e.g., in the mid-term by assigning differently sized aircraft to different routes (cf. Berge and Hopperstad 1993) or by buying or leasing additional aircraft (cf. Stonier 1999). Since the problem considered here is situated within the short-term planning phase, such approaches are beyond the scope of the thesis and are not considered in the following.

For short-term forecasting, the availability of historic and current booking data as well as data on the historic cancellation and no-show behavior of customers represent a prerequisite for forecasting demand by the targeted market segments (see above, Sec. 3.3.2). According to Swan (2002), the most common assumption about demand distributions in revenue management systems is the normal distribution although it technically exhibits always a chance of negative values which for demand in reality cannot exist. The author therefore suggests a combination of normal and Gamma models, which so far has not been proposed for use in revenue management optimization (ibid., p. 262). Swan (2002, p. 262) further notes that "bookings for total airplane loads [. . .] are either Normal or of such a high-order Gamma that Normal is close enough". Hence, for the model introduced in Chap. 4, the assumption of normally distributed demand is maintained.

3.3.3.2 Pricing

The tariff structure in a RM system is based on the principle of (second- and third-degree) price discrimination (cf. Pigou 1932; Talluri and van Ryzin 2004, p. 352 f.). Ideally, the seller first segments the market into groups of customers by their respective price sensitivity and creates restrictions between the segments that prevent dilution (see above, Sec. 3.3.2). Then, the seller establishes a price (fare) for each class of customers based on forecasted demand and allocates the fixed capacity among the fare classes. Over the booking period, the price for each fare class may remain constant, however, fare classes may become unavailable if the capacity allocated to a fare class is sold out. For the basic scenario with two market segments exhibiting two different linear demand curves, Reece and Sobel (2000) provide a graphic approach to determining the profit-maximizing price and capacity share for two fare classes. In a more complex setting, Ladany and Arbel

(1991) investigate the optimal market segmentation and pricing strategy for passenger cabins on cruise liners.

Taking into account varying arrival patterns among customer segments, Desiraju and Shugan (1999) derive strategic pricing recommendations and argue that, going beyond the basic segmentation and pricing approach illustrated by Reece and Sobel (2000), the success of a revenue management approach also may depend on dynamic pricing. In the classic revenue management approach, price variations result only from controlling the availability of fare classes by closing (and potentially re-opening) fare classes over the booking period. In contrast, dynamic pricing models the price for each class as a function of time and directly adjusts prices at critical points in time.

The first, classic approach – with a price menu and controlled by opening and closing of fare classes – is sometimes referred to as *quantity-based revenue management*, the second, dynamic-pricing approach as *price-based revenue management* (Talluri and van Ryzin 2004, p. 20 f.). Quantity-based RM usually requires a somehow differentiated product or service, respectively, for each fare class, e.g., by introducing service classes like first, business, and economy class or restrictions with regard to cancellation, changes in booking, or length of stay. Price-based RM, however, may also be applied for a homogenous, non-differentiated product as, e.g., offered by many low-cost airlines, in retailing, or manufacturing.

With regard to the required product differentiation in the case of quantity-based RM, Botimer and Belobaba (1999) provide a theoretical framework that explicitly takes into account the degrading cost incurred by consumers when accepting more restrictions and incorporates the diversion of passengers to lower-priced fare products.

Feng and Xiao (2000a, b) determine the optimal timing of a seller who offers a product at a predetermined set of multiple prices to change prices monotonically, i.e., markup or markdown, based on the remaining season and inventory. Similarly, Feng and Gallego (2000) investigate the optimal timing of price changes from a finite number of allowable price paths, each following a general Poisson process with Markovian time-dependent demand intensities. A comprehensive overview on pricing models for revenue management is provided by Bitran and Caldentey (2003).

In a multi-product environment, Maglaras and Meissner (2004) show that the two approaches of a dynamic pricing strategy for each product or of a dynamic rule to control capacity allocation if product prices are fixed can be reduced to a common formulation that can lead to algorithmic simplifications.

While RM usually assumes that the seller knows the demand distribution ex ante, Petruzzi and Dada (2002) relax this assumption and allow for the seller to learn about demand parameters only over the selling season and identify updates of the seller's subjective probability distribution of demand as a factor that influences the optimal selling price over the selling season.

3.3.3.3 Inventory Control

While the literature contributions introduced in the previous section are primarily concerned with determining optimal prices for individual fare classes, optimally selecting a price from a given price menu, or determining the optimal timing for a price change, the contributions to the literature reported on in the following usually take the set of fare products and prices as given. Instead, inventory control focuses on allocating capacity to fare classes, determining the optimal availability/non-availability policy for fare classes, and thus on whether or not to accept or denying booking requests.

Classic seat-inventory control models for passenger airlines assign a booking limit to each fare class that indicates the maximum number of seats to be sold in the respective fare class (cf. Belobaba 1987) in order to limit low-fare seats and protect seats for late booking customers in higher fare classes. For a two-tiered pricing strategy with a discount-fare and a full-fare class, Pfeifer (1989) presents an extended Newsboy model to determine the optimal booking limit. With a similar setting, Belobaba (1989) reports on the implementation of a computerized "Automated Booking Limit System" at Western Airlines in 1987.

Fare classes can be modeled either nested or non-nested. In non-nested systems, each fare class is separately assigned a number of seats. In nested systems, bookings for high-margin classes are always accepted as long as capacity is available in the high-margin class *or* in any lower-margin class. Low-fare classes are thus nested into the high-fare classes such that the highest fare class has access to the total seat inventory and lower fare classes to a subset only. For a system with multiple nested fare classes, Brumelle and McGill (1993) provide a model to determine the optimal seat allocation. For two nested fare classes, Feng and Xiao (2000c) present an alternative approach based on the theory of optimal stopping time for when to close the low-price fare class as to assure a certain service level in the high-price class.

An important limitation of the classic fare-class control models is that optimization is performed for each flight leg individually. However, an airline is usually concerned with maximizing the revenue of its total network, especially when operating a hub-and-spoke network, but maximizing revenue of each flight leg independently does not guarantee that total network revenue is maximized (cf. Williamson 1988) because many passenger itineraries involve more than one flight leg.

To overcome this limitation, airlines have developed leg-based origin and destination (O&D) control heuristics, which – though leg- and not network-based – tries to give priority to bookings with itineraries/fares that have the highest value contribution to the total network by redefining booking classes as value classes according to their network value contribution (Belobaba 1998). Due to their computational complexity, network-based optimization models have only recently become viable alternatives to such heuristics. For a system with multiple origins, one hub, and one destination, Feng and Xiao (2001) present a stochastic control model and develop optimal control rules. Also for a multi-leg flight and a single-hub network, de Boer et al. (2002) investigate the revenue impact of approximative, deterministic versus more advanced, probabilistic demand models.

Instead of static booking limits as in classic RM models, inventory control is performed in these settings by the implementation of bid prices. Bid prices are threshold values that represent the opportunity cost of selling one unit of capacity. A product, be it a seat in a fare class on a one- or multiple-leg itinerary or a hotel room for one or more consecutive nights, is only sold if the offered or requested fare exceeds the sum of the threshold values of all resources required for supplying the product, e.g., for all legs of an itinerary (cf. Talluri and van Ryzin 1998, 1999).

Bitran and Caldentey (2003), however, show that bid-price policies are generally suboptimal, but (in line with Cooper 2002) asymptotically optimal as capacity increases. Bertsimas and Popescu (2003) investigate the optimal capacity allocation in a network environment with multiple fare classes and use a new algorithm based on approximate dynamic programming for which they report that it leads to higher revenues and more robust performance than bid-price control.

Recent publications explicitly model the consumer behavior and its effects on demand for certain fare-type products. Zhao and Zheng (2001) show that demand is not only affected by a fare's current availability but also by the possibility of its availability in the future. Talluri (2004) specifies the probability of purchase as a function of the set of the fare menu offered by the seller and dynamically determines the optimal fare menu at each point in time.

A revenue management approach different from the traditional approach with price menu and capacity controls is presented by Vulcano et al. (2002). In a multi-period setting, a seller faces a random number of buyers per period. Each buyer bids in a dynamic auction for the capacity, thereby facing competition from other buyers in the same period and from buyers in subsequent periods through the opportunity cost of capacity assessed by the seller. Vulcano et al. (2002) show that the auction mechanism outperforms the traditional RM approach in the case of a moderate number of periods and moderate sales volume.

Anderson et al. (2004) likewise present a novel revenue management approach based on real options. In the context of a car rental firm, the authors model the firm's decision of accepting or not accepting a booking at a specific rate and point-in-time as the exercise decision of a "swing" option and determine that way both the firm's optimal car rental strategy and the value of the car rental business to the firm.

3.3.3.4 Overbooking

Managing capacity utilization in the airline industry is aggravated by cancellations and no-shows[6] of passengers. Smith et al. (1992, p. 11) state that, "on average, about half of all reservations made for a flight are canceled or become no-shows". Starting in the 1960s, airlines have countered cancellations and no-shows by overbooking, i.e., by accepting a higher number of reservations than seats are available. The crucial question is to what extent an airline should overbook each particular flight. On the one hand, empty seats mean an opportunity cost to the airline. For instance, American Airlines estimates that about 15% of seats would remain unused on flights sold-out at departure and an even higher percentage on flights sold-out before departure (Smith et al. 1992, p. 9). On the other hand, overbooking entails the risk of overselling, i.e., more passengers actually showing-up than seats are available, and thus compensation and loss-of-goodwill cost. Overbooking models essentially trade-off the cost of overbooking and under-utilized capacity (see Fig. 3.4).

The earliest formal overbooking models date back to Beckmann (1958) who determines in a static approach the optimal overbooking level in a single-fare-class setting. A general, non-industry-specific formulation of this setting and modeling approach present Bodily and Pfeifer (1992). In contrast to static approaches that determine the overbooking level once, dynamic models recalculate and adjust the overbooking level over the booking period. Though such approaches probably suit better the dynamic nature of the real-world problem, they are also more complex, which may hinder their real-world implementation (cf. Tscheulin and Lindenmeier 2003, p. 642 f.). For the single-fare-class setting, Chatwin (1998) presents a multi-period overbooking model, which is extended in Chatwin (1999) to allow for time-dependent fares. Chatwin (1996) develops an overbooking model for a multiple-fare-class setting. Also for many (substitutable) inventory classes but in a two-period setup, Karaesmen and van Ryzin (2004) show that the overbooking level in one reservation class decreases in the number of bookings for other reservation classes and derive joint optimal overbooking levels.

---

[6] A passenger who does not cancel his or her booking but does not show-up to honor the reservation at the day of departure is considered a no-show.

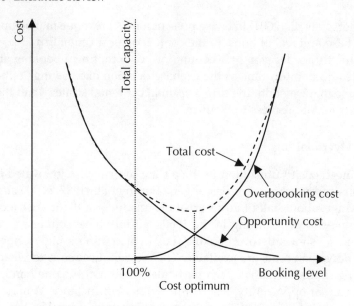

Fig. 3.4. Cost trade-off for overbooking (schematic). Overbooking cost include refunds, compensations, as well as cost of goodwill loss.

3.3.4 Implementation and Impact

Besides work published on the actual optimization components at work in a revenue management system, a part of the literature focuses on implementation issues and the impact of revenue management both on the seller's financial performance and the buyer's perception.

The financial benefits from the implementation of revenue management systems for passenger airlines are reported to amount to an increase of 3–7% in revenue and a 50–100% increase in profit (Smith et al. 1992; Cross 1998). According to Smith et al. (1992, p. 25), American Airlines increased revenues by $225 million only from overbooking in 1990.

Lahoti (2002b) focuses on the implementation of revenue management systems and stresses the necessity of management commitment for a successful implementation. Smith et al. (2001) describe the prospects of airline marketing and the opportunities for revenue management resulting from using internet platforms in business-to-business and business-to-consumer relationships. Going one step further, Toh and Raven (2003) propose an integrated internet marketing strategy that puts the airline's revenue management activities at the center while embracing all marketing and customer relationship management activities in order to archive RM goals in consistency with consumer needs.

Price

	Fixed	Variable
	Quadrant 1	*Quadrant 2*
Predictable	Movies Stadiums and arenas Convention centers Hotels' function space	Hotel rooms Airline seats Rental cars Cruise lines
	Quadrant 3	*Quadrant 4*
Unpredictable	Restaurants Golf courses	Continuing care Hospitals

Duration (left axis label)

Source: Kimes and Chase (1998, p. 157)

Fig. 3.5. Typical pricing and duration positioning of selected service industries

Consistency with consumer needs is an important issue because it has been shown by other authors that the application of revenue management techniques may be perceived as unfair by consumers. Kimes and Wirtz (2003) investigate in the context of restaurant revenue management the perceived fairness of demand-based pricing and point out that some price fences, e.g., price differentiation by table location, are perceived less fair than others, e.g., couponing and time-of-day pricing. Comparing the hotel and airline industry, Kimes (1994) finds that the acceptance of RM practices among airline passengers is higher than among hotel guest and attributes this finding to the longer history of RM in the airline industry. In a follow-up study, Kimes (2002) ascertains an equal level of customer perception of demand-based pricing policies in both industries. In addition, Kimes (1994) emphasizes that customers view information on pricing options as essential and that they are willing to accept reasonable restrictions on a product or service when they are offered a price discount in exchange.

3.3.5 Areas of Application

Because revenue management originated from the passenger airline industry, the majority of publications has developed in this field. However, revenue management principles can in general be applied in industries that show the characteristics enumerated in Sec. 3.3.2.

Depending on the nature of the service industry, though, different RM tools have proven more effective than others. Kimes and Chase (1998) identify two characteristics of a service offering – pricing flexibility and duration predictability – depending on which two strategic levers of revenue management – price management and duration (or inventory) control – should be applied to a greater or lesser extent. Industries that are apt to the application of RM can be classified on the basis of their use of the two strategic levers (see Fig. 3.5). Passenger airlines, as discussed above, exhibit a wide set of prices but offer a service whose duration, i.e., length of customer use, is clearly predictable and are thus located in Quadrant 2. The emphasis in the application of RM tools is thus on pricing and inventory control, as discussed in Sections 3.3.3.2 and 3.3.3.3. In contrast, the duration of the service offered by a restaurant is unpredictable and prices are usually fixed (Quadrant 3), making duration control, i.e., pacing and prediction of customer arrivals and length of stay, the predominant lever for revenue management.

Partly due to its similarity with the passenger airline industry (Quadrant 2), an area that has early caught the attention of RM researchers is the hotel and lodging industry. An analysis of RM systems in UK hotels is presented by Jones (1999). Kimes (2003) summarizes the genesis of RM in the hotel industry. In analogy to the above discussion of bid prices in the airline industry, Baker and Collier (2003) investigates bid-price methods, which they claim to represent the current standard in the hotel industry, and develop a price setting method that the authors report to be superior to bid prices.

Carroll and Grimes (1995) describe the introduction of a revenue management system at a car rental company. The significance of RM given the high-fixed-cost, utilization-driven structure of the car rental industry is underlined by Geraghty and Johnson (1997) who report on the turn-around of almost bankrupt National Car Rental driven by applying RM techniques.

Other areas, which RM has successfully been applied to, include restaurants (Kimes et al. 1998), contract manufacturing (Harris and Pinder 1995), broadcasting (advertising slots) (Cross 1997, p.150), telecommunication companies (Humair 2001), internet service providers (Nair and Bapna 2001), non-profit companies (Metters and Vargas 1999), golf courses (Kimes 2000), passenger railways (Ciancimino et al. 1999), and cruise liners (Ladany and Arbel 1991). The application of RM in the air cargo industry is discussed separately in the following section.

3.3.6 Air Cargo Revenue Management

The application of revenue management techniques is a relatively recent development in the air cargo industry: "Very little attention was paid to revenue management for cargo airlines until a few years ago," note Herrmann et al. (1998a, p. 392). Talluri and van Ryzin (2004, p. 563) describe the use of RM in air cargo at present as "rather sporadic". Holloway (2003, p. 572) cites the following reasons for the air cargo industry adopting RM relatively late: short booking cycles, multi-dimensional nature of freight shipments, widespread-use of negotiated rates (contracts), uncertainty of cargo capacity on passenger flights, and the fact that spot prices on routes with excess output of cargo space often undercut contract rates. Talluri and van Ryzin (2004, p. 563 f.) emphasize the long-term nature of customer relationships and the fact that the bulk of space is sold under long-term contracts to few important customers as reasons for air cargo RM lacking the sophistication of passenger airline RM.

The body of literature that specifically deals with revenue management in air cargo is rather small – especially with regard to scientific papers published in scholarly journals. The major contributions in the latter are briefly summarized in the following.

Kasilingam (1996) first tapped the field of air cargo revenue management by comparing the business of passenger airlines with cargo carriers and discussing the characteristics and complexities of revenue management in the airfreight business. According to Kasilingam, four important differences distinguish cargo and passenger revenue management: uncertainty of capacity, three-dimensionality of capacity (weight, volume, and number of container positions), itinerary control (cargo can be shipped on any route within a cargo carrier's network), and allotments. With regard to the practice of reserving allotments for big customers (major shippers and forwarders), the author notes that "this practice requires decision support to identify the flights for which allotments need to be set up and the amount of allotment space." (ibid., p. 38). Kasilingam also describes the cargo revenue management process and identifies four important steps:

1. The airline needs to forecast the cargo capacity available for sale (for cargo capacity on passenger aircraft).
2. Space for allotments is allocated, based on demand, allotment profitability, and anticipated spot market sales.
3. The remaining capacity (forecasted capacity less allotments) is overbooked to compensate for cancellations and no-shows.
4. The overbooked capacity is allocated to different market segments (buckets) and products so that total cargo revenue is maximized. (Ibid., p. 38.)

These generic steps are also reflected by the procedure at LCAG described in Sec. 2.2.3. The model and its extensions introduced in Chapters 4 and 6 will focus on the second and third step of this framework. With regard to the third and fourth step, Kasilingam (1996) introduces a model for air cargo overbooking with stochastic capacity and a bucket allocation model. This overbooking model is extended in Kasilingam (1997) to allow for continuous distributions of capacity and solved exemplary for a discrete distribution and the (continuous) uniform distribution.

More recently, Slager and Kapteijns (2004) have described the implementation of cargo revenue management at KLM Royal Dutch Airlines, starting in the mid-1990s. Like LCAG (see Sec. 2.2.3), KLM divides capacity sales into sales on the basis of contracts (termed "guaranteed capacity contracts" at KLM) and "free-sales" (termed "R/R") (ibid., pp. 83 f.). Overall, Slager and Kapteijns postulate a rather pragmatic approach to revenue management and the implementation of a basic set of processes and tools.

DeLain and O'Meara (2004) explore the expected return from an air cargo carrier's investment into a revenue management programme and identify the carrier's size (in terms of annual revenue) and route mixture as important criteria for the value of such an investment. For the air cargo carrier considered by DeLain and O'Meara, the expected total annual revenue increase from the application of revenue management techniques amounts to $5.3m or 3.2% of revenues.

The rather managerially than scientifically focused literature has just recently paid more and more attention to (air) cargo revenue management. With regard to the subject of this thesis, the following contributions are especially relevant:

Pompeo and Sapountzis (2002) note that the performance in the freight industry has consistently underperformed the S&P 500 in the second half of the 1990s. The authors argue that the introduction of "risk and revenue management could provide the boost the industry needs" (ibid., p. 90). With regard to market segmentation strategies like LCAG's product offering (see Sec. 2.2.1), Pompeo and Sapountzis (2002, p. 97) report that "one leading air cargo carrier increased its yield by about 5 percent from 1995 to 1998 while the average industry yield fell by 15 percent." The authors also stress the negative impact of contract anomalies (see also p. 18) and price uncertainty. Price swings create gaps between spot and contract prices that give customers an incentive to default from capacity contracts (ibid., p. 93). Pompeo and Sapountzis advocate spreading risk more evenly between shippers, forwarders, and carriers. Furthermore, they postulate carriers offering "a discount for forward contracts they can interrupt if spot prices soar" (ibid., pp. 94 f.).

Ott (2003) likewise advocates risk-sharing in the carrier-forwarder relationship: "Risk sharing is essential in this area of high economic pressure and war threats, if for no other reason than the long-term survivability of cargo operations" (ibid., p. 49). However, with regard to the trend to reserving space in advance to share capacity utilization risk, the author reports scepticism that forwarders will change "the practice of bidding for cargo space in the closing days and hours prior to a cargo aircraft departure" (ibid., p. 48) because of the high market power of forwarders.

In a survey article, Herrmann et al. (1998a) discuss the influencing parameters on air cargo revenue management, under which they subsume capacity contracts and pricing, and identify overbooking, fare mix, and upgrading as most important instruments.[7] Herrmann et al. (1998a, p. 399) identify demand, capacity, cost, and competition as primary factors that influence the price of airfreight capacity. With regard to market segmentation, Herrmann et al. argue that price and time sensitivity represent the main segmentation criteria. Customers who need to ship cargo suddenly and as fast as possible or shippers with high-value freight are usually price insensitive. Those who are able to foresee their need for air cargo and thus book well in advance are usually price sensitive (Herrmann et al. 1998a, p. 393).

Lahoti (2002a) discusses challenges and shortcomings of current revenue management systems in air cargo revenue management. As challenges Lahoti identifies among others "to allot space for long-term contracts vs. spot sales based on supply and demand", "to price different product configurations under long-term or spot conditions to maximize revenues based on capacity, demand, and price elasticity", and "to optimize overbooking to minimize service failure cost."

To conclude, the majority of the authors agree that though airfreight fulfills the criteria for a successful application of revenue management (see Sec. 3.3.2), that one of the important additional challenges for air cargo revenue management is constituted by allotments and the existence of long-term contracts. However, analytic models that incorporate this specific requirement are rare. This work aims at narrowing this gap in the literature. It tackles the issue of how to integrate allotments into the revenue management process and models the building of allotments for contract and spot market and pricing of capacity contracts. Thereby, the model presented in this work goes beyond current market practice by integrating capacity-option contracts into the revenue management process.

---

[7] The text at hand, in line with the academic literature, follows a more holistic line of definition and understands capacity contracts and pricing as integral parts of a company's revenue management system rather than influential parameters.

4

Capacity-Option Pricing Model

This chapter contains the formulation of a two-period model with an intermediary and an asset provider who interact with each other in a contract market and with other buyers and sellers in a subsequent spot market for airfreight capacity. First, in Sec. 4.1, an overview of the model formulation and setup is provided. Sec. 4.2 details and discusses the underlying assumptions. Then, the decision problems of the market participants are presented and optimal policies derived (Sec. 4.3 and 4.4).

4.1 Model Overview and Setup

The model considers one asset provider and one intermediary. The asset provider A offers transport capacity K on a specific route at a specific date in the future. The capacity is fixed, known, and divisible. Per unit of capacity, the asset provider incurs a fixed cost of f. He can sell the capacity or a portion of it upfront – on the contract market long before the date of transport – to the intermediary I and the remainder on the spot market shortly before the date of transport. If the asset provider sells on the spot market, he is paid the (uncertain) spot price $\tilde{s}$ per capacity unit and incurs a cost t.[1] If he sells on the contract market, he is paid a price of r and incurs a cost of c for every capacity unit that is reserved by the intermediary. If the intermediary actually calls on the reserved capacity on the day of transport, he pays the asset provider a price of x who then incurs a cost of v. Note that when $x = 0$, r can be thought of as the "wholesale" price (then denoted by w) and the contract between the intermediary and the asset provider as a fixed-commitment (forward) contract. When $x > 0$, the contract can be

---

[1] In the following, a tilde ($\tilde{}$) on a variable or function indicates a stochastic variable or a function of a stochastic variable.

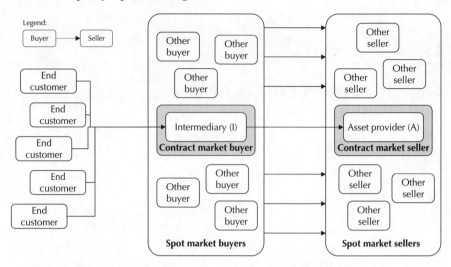

Fig. 4.1. Participants in contract and spot market

thought of as an option contract with a reservation fee of r and an execution fee of x.

The intermediary can procure capacity on the contract market from the asset provider or on the spot market from any other spot market seller. He bundles the capacity with additional services of his own and sells this service bundle to end customers at price p which is a function of his procurement cost and markup λ, i.e., $p = p(r, x, \lambda)$. He faces uncertain, price-dependent demand $\tilde{D}_C(p)$ from end customers. In the contract market, the asset provider thus faces demand $\tilde{D}_C$ from the intermediary. Demand $\tilde{D}_S$ from other buyers on the spot market that the asset provider faces is independent and uncertain (see below, Sec. 4.2.2), with $F_S(\tilde{D}_S)$ and $f_S(\tilde{D}_S)$ denoting the cumulative distribution function and distribution density function of $\tilde{D}_S$, respectively. (The market structure is depicted in Fig. 4.1.)

The sequence of events is the following (see Fig. 4.2): First, in the contract market phase, the asset provider announces r and x and the intermediary decides on the number of capacity units to reserve (denoted by N). The subsequent spot market phase is split into two parts, the booking and the execution phase. During the booking phase, the intermediary collects bookings from end-customers and can book additional capacity on the spot market. The asset provider offers up to B capacity units (in the following, B is referred to as the *booking limit*) on the spot market and receives up to $\tilde{M}$ bookings (with $\tilde{M} \leq B$). During the execution phase – at the date of transport, say –, demand $\tilde{D}_i$ ($i = C, S$) and spot price $\tilde{s}$ are realized and the intermediary informs the asset provider of the number of reserved capacity units he calls on, i.e., executes (denoted by $\tilde{E}$ with $\tilde{E} \leq N$). Likewise, spot

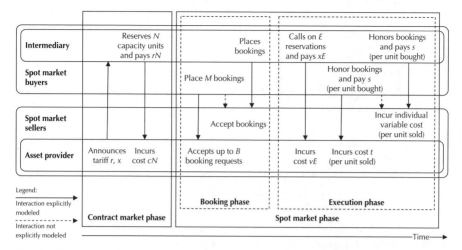

Fig. 4.2. Sequence of events in contract and spot market. Direction of arrows indicate direction of action/payment.

market buyers who have placed a booking with the asset provider during the booking phase show up to honor their bookings. (Table 4.1 summarizes the notation introduced so far).

For the time being, it is assumed that the asset provider only offers the non-reserved capacity in the spot market, i.e., $B = K - N$. This assumption is relaxed in Sec. 6.3, allowing for $B > K - N$ and thus overbooking of capacity.

4.2 Assumptions

4.2.1 Market Structure

The analysis in the contract market is limited to one seller and one buyer. However, competition by other sellers is included indirectly via the demand function (see below). The extension of the model to multiple buyers is straight forward: The current single buyer can be interpreted as the aggregation of all buyers without changing the seller's profit function.

The spot market is assumed to be perfectly competitive with a large number of buyers and sellers who individually act as price takers.

Asset provider and intermediary are risk neutral and maximize expected profits. The effect of discounted cash flows is ignored. These assumptions follow the vast majority of the operations research literature on revenue management (cf. Weatherford and Bodily 1992; Feng and Xiao 1999).

Table 4.1. Overview of model variables and functions introduced so far

Variable	Description
$\tilde{D}_C$	Contract market demand faced by intermediary and asset provider
$\tilde{D}_S$	Spot market demand faced by the asset provider
$\tilde{E}$	Number of called-on reservations
K	Total capacity
B	Maximum number of spot market bookings A is willing to accept (booking level)
$\tilde{M}$	Actual number of spot market bookings received by A (spot market sales)
N	Number of capacity units reserved by I
c	Cost incurred by A per reserved capacity unit (reservation cost)
f	Cost incurred by A per capacity unit (fixed cost)
r	Price for reserving one capacity unit (reservation fee)
$\tilde{s}$	Price for one capacity unit in the spot market (spot price)
t	Cost incurred by A per capacity unit sold in spot market (spot cost)
v	Cost incurred by A per called-on reserved capacity unit (execution cost)
w	Price for one capacity unit in the forward-contract market
x	Price for calling on one reserved capacity unit (execution fee)
λ	Intermediary's markup

4.2.2 Supply and Market Demands

The capacity supply in the spot market is unlimited, i.e., for the spot price of $\tilde{s}$ per capacity unit the intermediary can always buy any amount of capacity. Likewise, the asset provider can sell capacity at $\tilde{s}$ in the spot market, however only up to the amount of the minimum of demand $\tilde{D}_S$ (see below) and his total capacity K.

The demands in spot and contract market are independent. This assumption is a simplification that many revenue management models revert to (cf. Weatherford and Bodily 1992; Tscheulin and Lindenmeier 2003). However, with regard to the market characteristics and participants as introduced in Sec. 2.2 it can be justified to regard these markets as sufficiently separated because the market structure differs (see the preceding section) and the asset provider aims at serving different market segments with different price sensitivities.

The demand function in the contract market is

$$\tilde{D}_C(p) = a - bp + \tilde{\epsilon} \qquad (4.1)$$

with $a > 0$, $b \geq 0$ and with $\tilde{\epsilon}$ being normally distributed with mean $\mathcal{E}[\tilde{\epsilon}] = 0$ and standard deviation $\sigma_{\tilde{\epsilon}}$, i.e., $\tilde{\epsilon} \sim N(0, \sigma_{\tilde{\epsilon}}^2)$.[2] In the linear demand function in (4.1), the ordinate intercept a can be interpreted as the maximum market size, i.e., the demand that could be generated if the price were 0; slope $-b$ measures the rate of change in market demand with respect to a change in price; the stochastic error term $\tilde{\epsilon}$ introduces uncertainty into the demand function (cf. Mills 1959; Lau and Lau 1988).

It follows that $\tilde{D}_C$ is also normally distributed with mean and standard deviation

$$\mu_{\tilde{D}_C} = \mathcal{E}[a - bp + \tilde{\epsilon}] = a - bp \quad \text{and} \quad \sigma_{\tilde{D}_C} = \sigma_{\tilde{\epsilon}}, \tag{4.2}$$

respectively. The distribution density function $f_C(\tilde{D}_C)$ and cumulative distribution function $F_C(\tilde{D}_C)$ are thus functions of p.

The portion of spot market demand that the asset provider sees is denoted by $\tilde{D}_S$ and is normally distributed with mean $\mu_{\tilde{D}_S}$ and standard deviation $\sigma_{\tilde{D}_S}$. It is assumed that this portion of total demand is independent of the current spot price $\tilde{s}$ (see below, Sec. 4.2.3). This assumption is further discussed in Sec. 6.2.1.

Both in contract and spot market, demand represents the sum of demands from a large number of customers. In the contract market, these are the (limited) number of end customers of the intermediary, in the spot market the (very large) number of spot market buyers.[3] Following the argumentation in Porteus (1990, p. 613), aggregated demands then are, due to the Central Limit Theorem, approximately normally distributed. Though technically a normal distribution implies a small chance that demand is negative, demands are assumed to be nonnegative, so that $F(\tilde{D}_i) = 0$ for $\tilde{D}_i < 0$ with $i = C, S$. The probability of negative demand depends on the coefficient of variation $sigma_i / \mu_i$. By keeping σ_i / μ_i sufficiently small, the chance of negative demand can be kept negligibly small, e.g., for $sigma_i / \mu_i = 1/3$, the chance of negative demand is 0.135%. Lau (1997, p. 560) suggest $sigma_i / \mu_i < 0.3$ to keep the negative tail negligible. For the model at hand, let $\mu_i \geq 3\sigma_i$ (with $i = \tilde{D}_C, \tilde{D}_S, \tilde{s}$). The assumption of normally distributed demand is widely used in supply-contract (cf. Tsay et al. 1999) and revenue management models (cf. Swan 2002).

4.2.3 Prices and Costs

The asset provider sets the prices r, x in the contract market. The spot market price $\tilde{s}$, however, is influenced neither by the asset provider nor the inter-

---

[2] The expected value of a variable or function is denoted by $\mathcal{E}[\cdot]$.

[3] The assumption with regard to the number of spot market buyers can be thought of as reflecting the situation where the asset provider sells capacity (also) via a B2B marketplace.

mediary but is a random variable. The cumulative distribution function of $\tilde{s}$ is denoted by $G(\tilde{s})$, the probability density function by $g(\tilde{s})$. The spot price $\tilde{s}$ is assumed to be strictly positive, so that $G(\tilde{s}) = 0$ for $\tilde{s} \leq 0$, and follows a normal distribution with mean $\mu_{\tilde{s}}$ and standard deviation $\sigma_{\tilde{s}}$. Again, to keep the chance of a negative spot price under this type of distribution negligibly small, the coefficient of variation $\mu_{\tilde{s}}/\sigma_{\tilde{s}}$ needs to be sufficiently small (see above). The assumption of a stochastic spot price captures the price uncertainty in the airfreight market as described in Chap. 2 that stem from both macroeconomic dynamics (see Sec. 2.1.2) and from short-term developments on a microeconomic level, e.g., *ad hoc* deals and dynamic pricing (see Sec. 2.2.3).

To determine his resell price p, the intermediary adds to his capacity procurement cost a fixed[4] amount of λ (with $\lambda \geq 0$) which covers his cost for value-adding services and profit. For simplicity and clarity of the exposition, it is assumed that the intermediary prices his services at

$$p = r + x + \lambda. \tag{4.3}$$

The assumption here is that the majority of the intermediary's business happens also on the basis of long-term agreements, e.g., with the intermediary taking on the role of a third-party logistics (3PL) provider for his end customers (see Sec. 2.2). The simplest case of a 3PL contract is a price-only contract. It is assumed that the intermediary closes such contracts with his end-customers featuring the price p. Once a contract has been signed, the intermediary is bound to the contract and cannot turn down demand from end customers. If the demand from end customers exceeds the amount of capacity the intermediary has in turn reserved with the asset provider, he purchases additional capacity on the spot market. To keep the model within a reasonable scope, it is assumed that the intermediary does so at his own risk, i.e., p is a function of the prices within the capacity contract with the asset provider only and not of the spot price.

The variable costs c, v, t incurred by the asset provider are assumed nonnegative and given. The variable transport cost on the basis of a spot market purchase is at least as high as the total variable cost on the basis of a capacity contract, i.e., $t \geq c + v$. Note that the variable cost of accepting a reservation c will be (close to) zero for many real-life applications, however it is included to allow for analysis of the influence of the cost structure on the model solution.

The salvage value of excess reservations is zero, i.e., if the intermediary does not face sufficient demand to fill the reserved capacity he has placed

---

[4] Using a relative, e.g., percentage, markup ($p = (1 + \lambda)(r + x)$) does not structurally change the results of the model.

with the asset provider, the reservations expire at no value. By assumption, the possibility to resell reservations on the spot market is excluded.

To avoid economically non-meaningful results, it is necessary to assume that the execution fee is positive, i.e, $x \geq 0$. If one allowed for negative values of x, the buyer would (if $x < 0$) always exercise all reservations since he would earn $-x$ per unit without incurring any cost. The execution policy would thus be deterministic and, in the absence of any time value of money (see Sec. 4.2.1), would de facto equal a fixed-commitment contract with $w = r - (-x)$.

4.3 The Intermediary's Problem

The intermediary and asset provider interact in a Stackelberg-like game. First, the asset provider as the Stackelberg leader announces r, x, then the intermediary as the follower decides on the amount N of capacity to reserve. After demand and spot price uncertainty have resolved, the intermediary decides on the number of reservations to call on $\tilde{E}$. The structure of the following analysis is reverse: First, the two-stage problem of the intermediary is presented and solved in the first stage for the optimal consumption policy and then for the optimal reservation policy, contingent on r, x. Then (see Sec. 4.4), the asset provider's optimal pricing policy is derived, contingent on the intermediary's reservation policy.

The intermediary's objective in the first-stage problem is to maximize profit, denoted by $\tilde{P}_I$, by choosing the optimal amount $\tilde{E}$ of reservations to call on (exercise) given the realizations of contract market demand $\tilde{D}_C$ and spot price $\tilde{s}$. Since the intermediary by assumption (see Sec. 4.2.3) fills all demand, the decision on $\tilde{E}$ contains the decision on the number of spot market purchases ($\tilde{D}_C - \tilde{E}$ if $\tilde{D}_C > \tilde{E}$). The problem formulation is:

$$\max_{\tilde{E}} \tilde{P}_I = \max_{\tilde{E}} p\tilde{D}_C - rN - x\tilde{E} - \tilde{s}(\tilde{D}_C - \tilde{E})^+ \tag{4.4}$$

$$\text{s. t.} \quad N \geq \tilde{E}, \tag{4.5a}$$

$$\tilde{E} \leq \tilde{D}_C, \tag{4.5b}$$

$$\tilde{E} \geq 0. \tag{4.5c}$$

Lemma 4.1. *For $x > 0$, the number of reservations $\tilde{E}$ that are called on is:*

$$\tilde{E} = \begin{cases} \min(\tilde{D}_C, N) & \text{if } 0 < x \leq \tilde{s}, \\ 0 & \text{if } x > \tilde{s}. \end{cases} \tag{4.6}$$

Proof. The constrained problem in (4.4) can be solved by applying the Kuhn-Tucker theorem (cf. Bertsekas 1999, p. 310). The Lagrangian is:

$$\mathcal{L} = p\tilde{D}_C - rN - x\tilde{E} - \tilde{s}(\tilde{D}_C - \tilde{E}) + \ell_1(N - \tilde{E}) + \ell_2(\tilde{D}_C - \tilde{E}) + \ell_3\tilde{E} \quad (4.7)$$

The first order conditions are:

$$\frac{\partial \mathcal{L}}{\partial E} = -x + s - \ell_1 - \ell_2 + \ell_3 = 0, \quad (4.8\text{a})$$

$$\ell_1(N - \tilde{E}) = 0, \quad (4.8\text{b})$$

$$\ell_2(\tilde{D}_C - \tilde{E}) = 0, \quad (4.8\text{c})$$

$$\ell_3\tilde{E} = 0. \quad (4.8\text{d})$$

Furthermore, it must hold true that

$$\ell_1 \geq 0, \quad \ell_2 \geq 0, \quad \ell_3 \geq 0, \quad (4.8\text{e})$$

$$N - \tilde{E} \geq 0, \quad \tilde{D}_C - \tilde{E} \geq 0, \quad (4.8\text{f})$$

$$\tilde{E} \geq 0. \quad (4.8\text{g})$$

The case $\ell_1 = 0$, $\ell_2 = 0$, $\ell_3 > 0$ gives the solution $\tilde{E} = 0$ with $x > \tilde{s}$. The case $\ell_1 > 0$, $\ell_2 = 0$, $\ell_3 = 0$ yields $\tilde{E} = N$ with $\tilde{D}_C > N$ and $\tilde{s} > x$; $\ell_1 = 0$, $\ell_2 > 0$, $\ell_3 = 0$ yields $\tilde{E} = \tilde{D}_C$ with $\tilde{D}_C < N$ and $\tilde{s} > x$. $\qquad \square$

As by intuition, the intermediary will thus only honor his reservations if $s \geq x$, else he will shift his capacity procurement to the then cheaper spot market.

For the special case of a fixed-commitment contract, where $x = 0$, the intermediary is assumed to make use of as many reservations as needed, i.e., $\tilde{E} = \min(\tilde{D}_C, N)$ for $x = 0$.

The second-stage problem of the intermediary is to determine the optimal number of reservations N^*, given the prices r and x, but with uncertainty with regard to contract market demand $\tilde{D}_C$ and spot price $\tilde{s}$. The intermediary's objective is to maximize expected profit, for notational convenience denoted by $\Pi_I \equiv \mathcal{E}[\tilde{P}_I]$. The problem formulation is:

$$\max_N \Pi_I = \max_N \mathcal{E}\left[p\tilde{D}_C - rN - x\tilde{E} - \tilde{s}(\tilde{D}_C - \tilde{E})^+\right]. \quad (4.9)$$

The expression for Π_I in (4.9) can be rephrased using (4.6). Building expectations then results in:

$$\Pi_I = p\mathcal{E}\left[\tilde{D}_C\right] - rN - \mathcal{E}\left[\tilde{D}_C\right]\int_0^x \tilde{s}g(\tilde{s})d\tilde{s} - \int_x^\infty xg(\tilde{s})d\tilde{s}\int_0^N \tilde{D}_C f_C(\tilde{D}_C)d\tilde{D}_C$$

$$- \int_x^\infty xg(\tilde{s})d\tilde{s}\int_N^\infty Nf_C(\tilde{D}_C)d\tilde{D}_C - \int_x^\infty \tilde{s}g(\tilde{s})d\tilde{s}\int_N^\infty (\tilde{D}_C - N)f_C(\tilde{D}_C)d\tilde{D}_C$$

$$(4.10)$$

With $\int_x^\infty \tilde{s}g(\tilde{s})d\tilde{s} = \mathcal{E}[\tilde{s}] - \int_0^x \tilde{s}g(\tilde{s})d\tilde{s}$ (and accordingly for $f_C(\tilde{D}_C)$), this can be rewritten as:

$$\mathcal{E}[\Pi_I] = (p - \mathcal{E}[\tilde{s}])\mathcal{E}\left[\tilde{D}_C\right] - rN \qquad (4.11a)$$

$$+ \int_x^\infty (\tilde{s} - x)g(\tilde{s})d\tilde{s}\left[\int_0^N \tilde{D}_C f_C(\tilde{D}_C)d\tilde{D}_C + \int_N^\infty Nf_C(\tilde{D}_C)d\tilde{D}_C\right]$$

$$= (p - \mathcal{E}[\tilde{s}])\mathcal{E}\left[\tilde{D}_C\right] - rN + \mathcal{E}\left[(\tilde{s} - x)^+\right]\mathcal{E}\left[\min(\tilde{D}_C, N)\right] \qquad (4.11b)$$

Theorem 4.1. *The optimal number of reservations N^* purchased by the intermediary is:*

$$N^* = \begin{cases} F_C^{-1}\left[1 - \frac{r}{\mathcal{E}[(\tilde{s}-x)^+]}\right] & \text{if } \mathcal{E}[(\tilde{s} - x)^+] \geq r, \\ 0 & \text{else.} \end{cases} \qquad (4.12)$$

Proof. Differentiating (4.11a) with respect to N and applying Leibniz' rule yields:

$$\frac{\partial \Pi_I}{\partial N} = -r + [1 - F_C(N)]\int_x^\infty (\tilde{s} - x)g(\tilde{s})d\tilde{s} \qquad (4.13)$$

Rearranging terms and setting $\partial \Pi_I / \partial N = 0$ (first order condition) results in the following condition for the optimal number of reservations N^*:

$$F_C(N^*) = \frac{\int_x^\infty (\tilde{s} - x)g(\tilde{s})d\tilde{s} - r}{\int_x^\infty (\tilde{s} - x)g(\tilde{s})d\tilde{s}} = \frac{\mathcal{E}[(\tilde{s} - x)^+] - r}{\mathcal{E}[(\tilde{s} - x)^+]} \qquad (4.14)$$

Simplifying and taking the inverse function of F_C yields (4.12).

For N^* to maximize Π_I, it must be that $\partial^2 \Pi_I / \partial N^2 < 0$ (second order condition). With

$$\frac{\partial^2 \Pi_I}{\partial N^2} = -f_C(N)\int_x^\infty (\tilde{s} - x)g(\tilde{s})d\tilde{s} \qquad (4.15)$$

this holds true for $\int_x^\infty (s - x)g(s)ds > 0$ since $f_C(N) > 0$ by definition. $\qquad \square$

The result in (4.14) corresponds to the solution of the standard Newsboy model (see Sec. 3.2.1.1) with underage cost $c_u = \mathcal{E}[(\tilde{s} - x)^+] - r$ and overage cost $c_o = r$. In the following, $\mathcal{E}[(\tilde{s} - x)^+]$ will be referred to as the *expected spot market premium* because it represents the price premium the intermediary expects to pay when procuring capacity from the spot market instead of using reserved capacity. If the intermediary has reserved less capacity than the demand by his end customers turns out to be, he incurs a shortage cost which is equal to the expected spot market premium less the reservation fee. If he has reserved more capacity than actually needed, he loses the reservation fee r on each excess capacity unit.

The condition $\mathcal{E}[(\tilde{s} - x)^+] \geq r$ in (4.12) represents the intermediary's participation constraint in the contract market. Only if the expected spot market premium is greater than the out-of-pocket expense for the reservation fee, the intermediary will reserve capacity. From the intermediary's point of view, paying the reservation fee can thus be interpreted as a hedge against the uncertainty of the spot market premium.

In the standard Newsboy model, the probability $F_C(N^*)$ is called the buyer's optimal *service level*. Here, it is the – for the intermediary – optimal probability that the reserved capacity N^* accommodates the entire demand from end customers.

Corollary 4.1. *The intermediary's expected profit Π_I is concave in N.*

Proof. The second derivative of Π_I with respect to N as in (4.15) is negative as shown in the proof to Theorem 4.1   □

There exists a unique value N^* that maximizes the intermediary's expected profit. Deviating from N^* by ordering less or more than N^* will lead to lower expected profit.

In order to calculate N^*, define $z_N \equiv (N^* - \mu_{\tilde{D}_C})/\sigma_{\tilde{D}_C}$ and $z_x \equiv (x - \mu_{\tilde{s}})/\sigma_{\tilde{s}}$ (see also Sec. A.1). Let $\varphi(\cdot)$ and $\Phi(\cdot)$ denote the probability density and cumulative distribution function of the standard normal distribution, respectively, and $L(\cdot)$ denote the standard normal loss function as defined in (A.10). Then N^* can be calculated by reformulating (4.12) as

$$N^* = \mu_{\tilde{D}_C} + z_N\sigma_{\tilde{D}_C} \quad \text{with} \quad z_N = \Phi^{-1}\left[1 - \frac{r}{\sigma_{\tilde{s}}L(z_x)}\right] \qquad (4.16)$$

Lemma 4.2. *N^* is decreasing in r and x.*

Proof. Let $N^* > 0$. For the lemma to hold, the partial derivatives of N^* with respect to r and x must be < 0. Because of (4.2) and (4.3) it is $F(\tilde{D}_C) = F(\tilde{D}_C(r, x))$. Differentiating (4.16) and applying the relation in (A.7) yields:

$$\frac{\partial N^*}{\partial r} = -\left[\frac{\sigma_{\tilde{D}_C}}{\varphi(z_N)\sigma_{\tilde{s}}L(z_x)} + b\right] \tag{4.17a}$$

$$= -\left[\frac{1}{f_C(N^*)\mathcal{E}[(\tilde{s}-x)^+]} + b\right] < 0 \tag{4.17b}$$

and

$$\frac{\partial N^*}{\partial x} = -\left[\frac{\sigma_{\tilde{D}_C}[1-\Phi(z_N)][1-\Phi(z_x)]}{\varphi(z_N)\sigma_{\tilde{s}}L(z_x)} + b\right] \tag{4.18a}$$

$$= -\left[\frac{[1-F_C(N^*)][1-G(x)]}{f_C(N^*)\mathcal{E}[(\tilde{s}-x)^+]} + b\right] < 0 \tag{4.18b}$$

Since $b > 0$ by definition, it follows that $-b < 0$. Since $f_C(N^*) > 0$ by definition and $\mathcal{E}[(\tilde{s}-x)^+] > 0$ if $r > 0$, the denominator in (4.17b) and (4.18b) is positive. Since the numerator in (4.17b) is strictly negative, it follows that $\partial N^*/\partial r < 0$. In (4.18b), it is $1 - F_C(N^*) > 0$ and $1 - G(x) > 0$, thus the numerator is negative and it follows that $\partial N^*/\partial x < 0$.   □

4.4 The Asset Provider's Problem

The asset provider's profit is denoted by $\tilde{P}_A$, the expected profit is $\mathcal{E}[\tilde{P}_A] = \Pi_A$. His decision problem is to choose the optimal combination of (r,x) that maximizes total expected profit under uncertainty with regard to the spot price as well as demand in contract and spot market. The formulation of the optimization problem is:

$$\max_{r,x} \Pi_A = \max_{r,x} \mathcal{E}\left[(r-c)N + (x-v)\tilde{E} + (\tilde{s}-t)\tilde{M} - fK\right] \tag{4.19}$$

$$\text{s. t.} \quad r \geq 0, \ x \geq 0.$$

The asset provider's profit is thus composed of the margin from selling reservations in the contract market $(r-c)N$, the margin from executed reservations $(x-v)\tilde{E}$, and the margin[5] earned on spot market sales $(\tilde{s}-t)\tilde{M}$, less the fixed capacity cost fK. N is taken from (4.12), $\tilde{E}$ from (4.6). $\tilde{M}$ is defined as:

$$\tilde{M} \equiv \min(B, \tilde{D}_S) \quad \text{with} \quad B = K - N. \tag{4.20}$$

---

[5] Formally, the formulation $(\tilde{s}-t)\tilde{M}$ includes the possibility that the asset provider earns a negative margin if $\tilde{s} < t$. This might be the case if, e.g., the asset provider is committed to offering scheduled service which he, for strategic reasons, does not cancel even if the spot price falls below variable cost.

Building expectations over (4.19) gives:

$$\Pi_A = (r - c)N$$

$$+ (x - v) \int_x^\infty g(\tilde{s}) d\tilde{s} \left[\int_0^N \tilde{D}_C f_C(\tilde{D}_C) d\tilde{D}_C + \int_N^\infty (N) f_C(\tilde{D}_C) d\tilde{D}_C \right]$$

$$+ (\mathcal{E}[\tilde{s}] - t) \left[\int_0^{K-N} \tilde{D}_S f_S(\tilde{D}_S) d\tilde{D}_S + \int_{K-N}^\infty (K - N) f_S(\tilde{D}_S) d\tilde{D}_S \right] - fK \quad (4.21)$$

For the distribution functions specified in Sec. 4.2 (normal distribution), this can be reformulated by applying the definitions in Sec. A.1 of the Appendix as:

$$\Pi_A = (r - c)N + (x - v)(1 - \Phi(z_x))(\mu_{\tilde{D}_C} - \sigma_{\tilde{D}_C} L(z_N))$$

$$+ (\mu_{\tilde{s}} - t)(\mu_{\tilde{D}_S} - \sigma_{\tilde{D}_S} L(z_B)) - fK \quad (4.22)$$

with $z_B \equiv (B - \mu_{\tilde{D}_S}) / \sigma_{\tilde{D}_S}$.

Theorem 4.2. *The asset provider's optimal tariff (r^*, x^*) satisfies the set of equations*

$$N - \left[r^* - c - (\mu_{\tilde{s}} - t)(1 - \Phi(z_B)) \right] \left[\frac{\sigma_{\tilde{D}_C}}{\varphi(z_N)\sigma_{\tilde{s}} L(z_{x^*})} + b \right]$$

$$- (x^* - v)(1 - \Phi(z_{x^*})) \left[\frac{\sigma_{\tilde{D}_C}(1 - \Phi(z_N))}{\varphi(z_N)\sigma_{\tilde{s}} L(z_{x^*})} + b \right] + \ell_1 = 0 \quad (4.23)$$

$$\left[\mu_{\tilde{D}_C} - \sigma_{\tilde{D}_C} L(z_N) \right] \left[(1 - \Phi(z_{x^*})) - (x^* - v)\frac{\varphi(z_{x^*})}{\sigma_{\tilde{s}}} \right]$$

$$- \left[r^* - c - (\mu_{\tilde{s}} - t)(1 - \Phi(z_B)) \right] \left[\frac{\sigma_{\tilde{D}_C}(1 - \Phi(z_N))(1 - \Phi(z_{x^*}))}{\varphi(z_N)\sigma_{\tilde{s}} L(z_{x^*})} + b \right]$$

$$- (x^* - v)(1 - \Phi(z_{x^*})) \left[\frac{\sigma_{\tilde{D}_C}(1 - \Phi(z_N))^2(1 - \Phi(z_{x^*}))}{\varphi(z_N)\sigma_{\tilde{s}} L(z_{x^*})} + b \right] + \ell_2 = 0$$

$$(4.24)$$

$$\ell_1 r^* = 0 \quad (4.25)$$

$$\ell_2 x^* = 0 \quad (4.26)$$

with

$$\ell_1, \ell_2, r^*, x^* \geq 0 \tag{4.27}$$

Proof. The constrained problem in (4.19) can be solved by applying the Kuhn-Tucker theorem (cf. Bertsekas 1999, p. 310). The Lagrangian is:

$$\mathcal{L} = \Pi_A + \ell_1 r + \ell_2 x \tag{4.28}$$

The first-order conditions are:

$$\frac{\partial \mathcal{L}}{\partial r} = \frac{\partial \Pi_A}{\partial r} + \ell_1 = 0, \tag{4.29a}$$

$$\frac{\partial \mathcal{L}}{\partial x} = \frac{\partial \Pi_A}{\partial x} + \ell_2 = 0, \tag{4.29b}$$

$$\ell_1 r = 0, \tag{4.29c}$$

$$\ell_2 x = 0. \tag{4.29d}$$

Furthermore, it must hold true that

$$\ell_1 \geq 0, \ \ell_2 \geq 0, \tag{4.29e}$$

$$r \geq 0, \ x \geq 0. \tag{4.29f}$$

The partial derivative of (4.19) with respect to r is:

$$\frac{\partial \Pi_A}{\partial r} = \frac{\partial[(r-c)N]}{\partial r} + (x-v)\frac{\partial \mathcal{E}\left[\tilde{E}\right]}{\partial r} + (\mathcal{E}[\tilde{s}] - t)\frac{\partial \mathcal{E}\left[\tilde{M}\right]}{\partial r} \tag{4.30}$$

Using the formulation in (4.22), the partial derivatives of $(r-c)N$, $\mathcal{E}\left[\tilde{E}\right]$, and $\mathcal{E}\left[\tilde{M}\right]$ with respect to r can be shown to be:

$$\frac{\partial[(r-c)N]}{\partial r} = N + (r-c)\frac{\partial N}{\partial r} \tag{4.31}$$

$$\frac{\partial \mathcal{E}\left[\tilde{E}\right]}{\partial r} = -(1 - \Phi(z_x)) \left[\frac{\sigma_{\tilde{D}_C}(1 - \Phi(z_N))}{\varphi(z_N)\sigma_{\tilde{s}}L(z_x)} + b \right] \tag{4.32}$$

$$\frac{\partial \mathcal{E}\left[\tilde{M}\right]}{\partial r} = -(1 - \Phi(z_B))\frac{\partial N}{\partial r} \tag{4.33}$$

Substituting (4.31)–(4.33) into (4.30) and (4.30) into (4.29a) yields (4.23).
The partial derivative of (4.21) with respect to x is:

$$\frac{\partial \Pi_A}{\partial x} = (r-c)\frac{\partial N}{\partial x} + \frac{\partial[(x-v)\mathcal{E}\left[\tilde{E}\right]]}{\partial x} + (\mathcal{E}[\tilde{s}] - t)\frac{\partial \mathcal{E}\left[\tilde{M}\right]}{\partial x} \tag{4.34}$$

Again using the formulation in (4.22), the partial derivatives of $(x - v)\mathcal{E}\left[\tilde{E}\right]$ and $\mathcal{E}\left[\tilde{M}\right]$ with respect to x can be shown to be:

$$\frac{\partial[(x - v)\mathcal{E}\left[\tilde{E}\right]]}{\partial x} = (\mu_{\tilde{D}_C} - \sigma_{\tilde{D}_C}L(z_N))\left[(1 - \Phi(z_x)) - (x - v)\frac{\varphi(z_x)}{\sigma_{\tilde{s}}}\right]$$

$$- (x - v)(1 - \Phi(z_x))\left[b + \frac{\sigma_{\tilde{D}_C}(1 - \Phi(z_N))^2(1 - \Phi(z_x))}{\varphi(z_N)\sigma_{\tilde{s}}L(z_x)}\right] \quad (4.35)$$

$$\frac{\partial\mathcal{E}\left[\tilde{M}\right]}{\partial r} = -(\mu_{\tilde{s}} - t)(1 - \Phi(z_B))\frac{\partial N}{\partial x} \quad (4.36)$$

Substituting (4.35) and (4.36) into (4.34) and (4.34) into (4.29b) yields (4.24).
□

Four cases for the optimal solution are possible. In the first case, it is $\ell_1 = 0$ and $\ell_2 > 0$, which results in a candidate solution for the optimal tariff with $r^* > 0$ and $x^* = 0$. The optimal contract would then be a fixed-commitment contract. In the second case, it is $\ell_1 = 0$ and $\ell_2 = 0$ and the decision variables of the candidate solution can take on positive values, i.e., $r^* > 0$ and $x^* > 0$. Then the optimal contract would be an option contract. In the third case, it is $\ell_1 > 0$ and $\ell_2 = 0$ and thus $r^* = 0$ and $x^* > 0$. The asset provider would de facto act as a pure spot market provider that accepts (priceless) pre-bookings in the contract market phase. The forth case, $\ell_1 > 0$ and $\ell_2 > 0$, is only theoretical and not economically meaningful since it implies $r^* = 0$ and $x^* = 0$. The shape of the asset provider's objective function in the first and second case is illustrated in Sec. 5.2.2 (see Fig. 5.4 and 5.5).

Given the form of (4.23) and (4.24), the optimal tariff (r^*, x^*) cannot be found analytically but requires recourse to numerical solution methods (see Sec. 5.1.3).

Theorem 4.3. Π_A *is jointly concave in r and x.*

Proof. For Π_A being jointly concave in r and x, it is necessary and sufficient that $\partial^2\Pi_A/\partial r^2 \leq 0$ and $\partial^2\Pi_A/\partial x^2 \leq 0$. However, given the form of the second derivatives in (B.4) and (B.8), this cannot be shown analytically. Instead it can be shown that the second derivatives of $\tilde{P}_A$ with respect to r and x are negative for all possible states of the world.

Starting with (4.4), initially two cases can be distinguished: $x > \tilde{s}$ and $x \leq \tilde{s}$.

Case 1 $x > \tilde{s}$
It follows from Lemma 4.1 that $\tilde{E} = 0$ and thus:

$$\max_N \tilde{P}_I = \max_N \tilde{D}_C(p - \tilde{s}) - rN \quad (4.37)$$

In this state, it follows that $N^* = 0$.

If $x \leq \tilde{s}$, two further cases can be distinguished: $\tilde{D}_C \leq N$ and $\tilde{D}_C > N$.

Case 2 $x \leq \tilde{s}$ and $\tilde{D}_C \leq N$
It follows from Lemma 4.1 that $\tilde{E} = \tilde{D}_C$ and thus:

$$\max_N \tilde{P}_I = \max_N \tilde{D}_C(p - x) - rN \tag{4.38}$$

In this state, it follows that $N^* = 0$.

For $x \leq \tilde{s}$ and $\tilde{D}_C > N$, it follows from Lemma 4.1 that $\tilde{E} = N$ and thus:

$$\max_N \tilde{P}_I = \max_N \tilde{D}_C(p - \tilde{s}) - N(r + x - \tilde{s}) \tag{4.39}$$

Two state-dependent solutions, for the cases $r + x - \tilde{s} > 0$ and $r + x - \tilde{s} \leq 0$, can be found for (4.39).

Case 3 $x \leq \tilde{s}$ and $\tilde{D}_C > N$ and $r + x - \tilde{s} > 0$
In this state, it follows from (4.39) that $N^* = \tilde{D}_C$.
Case 4 $x \leq \tilde{s}$ and $\tilde{D}_C > N$ and $r + x - \tilde{s} \leq 0$
In this state, it follows from (4.39) that $N^* = 0$.

Starting from (4.19) with the definition in (4.20), the asset provider's profit function can be written as:

$$\tilde{P}_A = (r - c)N^* + (x - v)\tilde{E} + (\tilde{s} - t)\min(K - N^*, \tilde{D}_S) - fK \tag{4.40}$$

The second partial derivatives of $\tilde{P}_A$ with respect to r and x can be determined for all possible states:

In Case 1 $x > \tilde{s}$ with $E = 0$ and $N^* = 0$
In this state, (4.40) simplifies to:

$$\tilde{P}_A = (\tilde{s} - t)\min(K, \tilde{D}_S) - fK \tag{4.41}$$

$$\frac{\partial \tilde{P}_A}{\partial r} = \frac{\partial \tilde{P}_A}{\partial x} = 0 \tag{4.42}$$

$$\frac{\partial^2 \tilde{P}_A}{\partial r^2} = \frac{\partial^2 \tilde{P}_A}{\partial x^2} = 0 \tag{4.43}$$

In Case 2 $x \leq \tilde{s}$ and $\tilde{D}_C \leq N$ with $E = \tilde{D}_C$ and $N^* = 0$
In this state, (4.40) simplifies to:

$$\tilde{P}_A = (x - v)\tilde{D}_C + (\tilde{s} - t)\min(K, \tilde{D}_S) - fK \qquad (4.44)$$

$$\frac{\partial \tilde{P}_A}{\partial r} = -(x - v)b \qquad \frac{\partial \tilde{P}_A}{\partial x} = \tilde{D}_C - (x - v)b \qquad (4.45)$$

$$\frac{\partial^2 \tilde{P}_A}{\partial r^2} = 0 \qquad \frac{\partial^2 \tilde{P}_A}{\partial x^2} = -2b < 0 \qquad (4.46)$$

In Case 3 $x \le \tilde{s}$ and $\tilde{D}_C > N$ and $r + x - \tilde{s} > 0$ with $E = N^*$ and $N^* = 0$

In this state, (4.40) simplifies to:

$$\tilde{P}_A = (\tilde{s} - t)\min(K, \tilde{D}_S) - fK \qquad (4.47)$$

$$\frac{\partial \tilde{P}_A}{\partial r} = \frac{\partial \tilde{P}_A}{\partial x} = 0 \qquad (4.48)$$

$$\frac{\partial^2 \tilde{P}_A}{\partial r^2} = \frac{\partial^2 \tilde{P}_A}{\partial x^2} = 0 \qquad (4.49)$$

In Case 4 $x \le \tilde{s}$ and $\tilde{D}_C > N$ and $r + x - \tilde{s} \le 0$ with $E = N^*$ and $N^* = \tilde{D}_C$

In this state, (4.40) requires distinguishing two further cases: $K - \tilde{D}_C > \tilde{D}_S$ and $K - \tilde{D}_C \le \tilde{D}_S$. For $K - \tilde{D}_C > \tilde{D}_S$, (4.40) can be written as:

$$\tilde{P}_A = (r - c - +x - v)\tilde{D}_C + (\tilde{s} - t)\tilde{D}_S - fK \qquad (4.50)$$

$$\frac{\partial \tilde{P}_A}{\partial r} = \frac{\partial \tilde{P}_A}{\partial x} = \tilde{D}_C - (r - c + x - v)b \qquad (4.51)$$

$$\frac{\partial^2 \tilde{P}_A}{\partial r^2} = \frac{\partial^2 \tilde{P}_A}{\partial x^2} = -2b < 0 \qquad (4.52)$$

For $K - \tilde{D}_C \le \tilde{D}_S$, (4.40) can be written as:

$$\tilde{P}_A = (r - c - +x - v - \tilde{s} + t)\tilde{D}_C + (\tilde{s} - t - f)K \qquad (4.53)$$

$$\frac{\partial \tilde{P}_A}{\partial r} = \frac{\partial \tilde{P}_A}{\partial x} = \tilde{D}_C - (r - c + x - v - \tilde{s} + t)b \qquad (4.54)$$

$$\frac{\partial^2 \tilde{P}_A}{\partial r^2} = \frac{\partial^2 \tilde{P}_A}{\partial x^2} = -2b < 0 \qquad (4.55)$$

Since $\Pi_A = \mathcal{E}[\tilde{P}_A]$ is by definition a probability-weighted linear combination (average) of $\tilde{P}_A$ in all possible states and since $\tilde{P}_A$ is jointly concave in r and x in all possible states, so is Π_A. $\qquad \square$

With Π_A being jointly concave in r and x, it is ensured that a candidate solution found for the system of equations in Theorem 4.2, which satisfies all constraints formulated in (4.27), is the global maximum Π_A.

5

Model Results and Comparative Statics

Having presented the structure of the capacity contract model and outlined the optimization problems of the market participants in the previous chapter, this chapter aims at analyzing the model with regard to the optimal type of contract, the optimal policies of buyer and seller, the market performance, and how these are influenced by environmental conditions, which are represented by the exogenous variables of the model.

The chapter is structured as follows: At first, the framework of the largely numerical analysis is set by defining the base case and three scenarios to be discussed. It follows an illustration of the optimal policies of buyer and seller. Then the benefit of the contract market is demonstrated by comparing a pure-spot market scenario with a scenario featuring a spot and (fixed-commitment) contract market. Subsequently, fixed-commitment and capacity option contracts are compared in an extensive comparative static analysis. Finally, the ability of capacity-option contracts to reduce double marginalization is investigated. The last section summarizes the findings from the chapter.

5.1 Framework for the Analysis

5.1.1 The Base Case

The analyses in this chapter will be conducted starting from a base case for which the values of the exogenous variables of the model are assumed as specified in Table 5.1.

Table 5.1. Values for exogenous variables in the base case

Variable	Value	Description
K	400	Capacity
a	300	Contract market demand function: ordinate intercept
b	8	Contract market demand function: slope
$\sigma_{\tilde{D}_C}$	20	Standard deviation of contract market demand
$\mu_{\tilde{D}_S}$	200	Mean of spot market demand
$\sigma_{\tilde{D}_S}$	20	Standard deviation of spot market demand
λ	8	Intermediary's markup
$\mu_{\tilde{s}}$	20	Mean of spot price
$\sigma_{\tilde{s}}$	5	Standard deviation of spot price
c	0	Variable reservation cost
v	5	Variable cost in contract market
t	5	Variable cost in spot market
f	8	Fixed capacity cost

The base case features the following properties that are chosen both to represent – in a stylized way – the situation described in Sec. 2.2[1] and to make the base case amenable to further analysis:

- Capacity K is set to a level that is of the order of magnitude (ca. $\pm 30\%$) of the sum of average market demands. With $b > 0$, the expected demand in the contract market $\mu_{\tilde{D}_C}$ is always smaller than the maximum theoretic contract market size a. The expected demand in the spot market is given by $\mu_{\tilde{D}_S}$.
- The variable cost t of providing capacity on a short-term basis (in the spot market) is initially assumed to be identical to the total variable cost $c + v$ of providing capacity on a long-term basis (in the contract market), i.e., $t = c + v$.
- The asset provider does not incur any cost for taking reservations, i.e, $c = 0$, but only a variable cost when reserved capacity is actually called on.
- Reflecting the high proportion of fixed capacity cost, the fixed cost f per capacity unit is higher than the variable costs t or $c + v$, respectively.

5.1.2 Scenario Definition

The following notation is introduced to distinguish three different scenarios:

---

[1] The base case reflects reasonable assumptions with regard to the proportion of the variables only, i.e., their *relative* orders of magnitude. Data giving information about the *absolute* order of magnitude in a real application is provided in Chap. 7.

Pure spot sales, indicated by superscript S. Under this scenario, the asset provider does not offer any capacity in the contract market. Since both parties are deprived of their decision variables under this scenario, the decision problems of intermediary and asset provider vanish and the expected-profit functions simplify to:

$$\Pi_I^S = \mathcal{E}\left[(p - \tilde{s})\tilde{D}_C\right] \quad \text{with} \quad p = \mu_{\tilde{s}} + \lambda \tag{5.1}$$

$$\Pi_A^S = \mathcal{E}\left[(\tilde{s} - t)\tilde{D}_S - fK\right] \tag{5.2}$$

The scenario serves as a lower limit to the expected profits earned by the parties and determines therefore the willingness of the parties to participate in the contract market. Since it will be shown that under the majority of realistic parameter constellations both parties do benefit from participating in the contract market, this scenario will only be reverted to in the case if one of the parties does not benefit.

Fixed-commitment contract plus spot sales, indicated by superscript FS. Under this scenario, the asset provider offers a fixed-commitment contract to the intermediary during the contract market phase *and* any remainder of capacity during the spot-market booking phase. The fixed-commitment contract is characterized by the absence of an execution fee, i.e., $x = 0$. The decision problem of the asset provider reduces to optimally setting the reservation fee. To distinguish the reservation fee set under this regime from the reservation fee when $x > 0$, it has been defined in Chap. 4 that $r \equiv w$ for $x = 0$. Note that this scenario is contained by the model formulation in Chap. 4 since the fixed-commitment contract is just a special case of the capacity-option contract: The optimization problem of the intermediary and its solution remain unchanged. The optimization problem of the asset provider simplifies to:

$$\max_w \Pi_A^{FS} = \max_w \mathcal{E}\left[(w - c)N - v\tilde{E} + (\tilde{s} - t)\tilde{M} - fK\right] \tag{5.3}$$

The optimality conditions reduce to

$$N - \left[w^* - c - (\mu_{\tilde{s}} - t)(1 - \Phi(z_B))\right]\left[b - \frac{\sigma_{\tilde{D}_C}}{\varphi(z_N)\mu_{\tilde{s}}}\right]$$

$$+ v\left[b - \frac{\sigma_{\tilde{D}_C}(1 - \Phi(z_N))}{\varphi(z_N)\mu_{\tilde{s}}}\right] = 0 \tag{5.4}$$

which is equivalent to (4.23) with $r^* = w^*$, $x = 0$, $\ell_1 = 0$, and, if $\tilde{s}$ is assumed strictly positive (see Sec. 4.2.3), $\lim_{x \to 0} L(z_x) = z_x = -\mu_{\tilde{s}}/\sigma_{\tilde{s}}$.

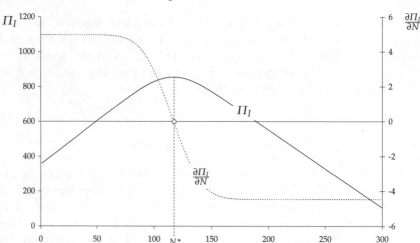

Fig. 5.1. Expected profit of intermediary Π_I as a function of the number of reservations when $r = 4.46$ and $x = 10.62$.

Capacity-option contract plus spot sales, indicated by superscript OS. Under this scenario, the asset provider offers a capacity-option or fixed-commitment contract (depending on the solution to the asset provider's optimization problem) to the intermediary during the contract market phase *and* offers any remainder of capacity during the spot-market booking phase.

5.1.3 Numerical Solution Procedure

Numerical solutions for prices w^*, r^*, and x^* presented in the following have been determined by using the spreadsheet solver software *What'sBest!*, Version 7 by *Lindo Systems Inc.*, Chicago (Ill.) in combination with the spreadsheet software *Microsoft Excel*, Version 10 by *Microsoft Corp.*, Redmond (Wash.).

5.2 Illustration of Optimal Policies

5.2.1 Optimal Reservation Policy of the Intermediary

The optimal reservation policy of the intermediary has been derived in Theorem 4.1. The intermediary chooses the number of reservations N such that his expected profits are maximized. Fig. 5.1 displays the objective function of the intermediary when the tariff is set by the asset provider to $r = 4.46$

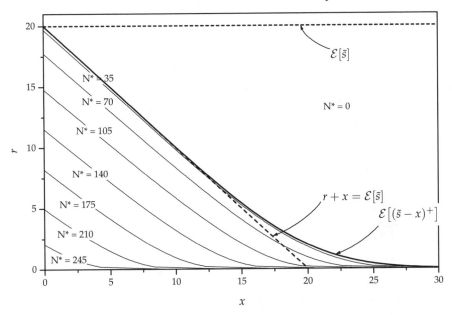

Fig. 5.2. Optimal reservation policy of the intermediary: Iso-quantity plot of the optimal number of reserved capacity units N^* in the (r, x) space.

and $x = 10.62$, which represents the optimal tariff under the set of parameters in Table 5.1. The expected profit Π_I is a function of the number N of capacity units reserved. As stated in Corollary 4.1, Π_I is concave in N and reaches its maximum at N^*.

The intermediary does not necessarily have to reserve capacity to conduct business. Under the set of parameters underlying Fig. 5.1, the intermediary also earns a positive profit if $N = 0$. He then procures its entire demand for capacity on the spot market from spot capacity providers. However, this procurement strategy is less profitable than reserving capacity with the asset provider since the expected cost to the intermediary per reserved capacity unit is smaller than the expected cost for spot market procurement.

This observation leads to the participation condition of the intermediary in the contract market as expressed in Theorem 4.1. As illustrated in Fig. 5.2, left and below the line marked $\mathcal{E}[(\tilde{s} - x)^+]$, the reservation fee r is smaller than the expected spot market premium $\mathcal{E}[(\tilde{s} - x)^+]$, thus the intermediary benefits from participating in the contract market and reserves capacity. Right and above this line, where $r > \mathcal{E}[(\tilde{s} - x)^+]$, no capacity is reserved.

Because the intermediary does in expectation not exercise all options and thus does not expect to actually pay $r + x$, it can be optimal for the intermediary to reserve capacity even if the sum of reservation and execution

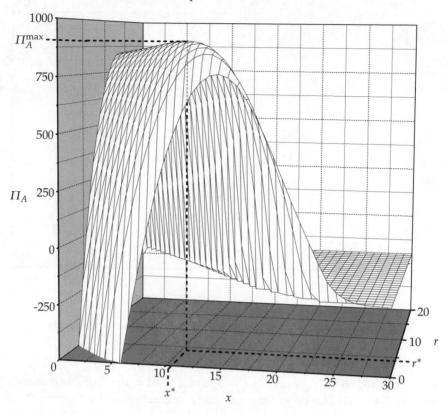

Fig. 5.3. Shape of the asset provider's objective function: Expected profit of asset provider Π_A as a function of reservation fee r and execution fee x.

fee exceeds the expected spot price $\mathcal{E}[\tilde{s}]$. In Fig. 5.2, this is the case for the area between the dashed bold diagonal indicating $r + x = \mathcal{E}[\tilde{s}]$ and the solid bold curve indicating $\mathcal{E}[(\tilde{s} - x)^+]$.

Fig. 5.2 also shows that the optimal number of reserved capacity units N^* is – as one would intuitively expect – decreasing in the prices r and x (see Lemma 4.2).

5.2.2 Optimal Pricing Policy of the Asset Provider

The optimal pricing policy of the asset provider has been formulated in Theorem 4.2. For the base case outlined in Table 5.1, the objective function of the asset provider can be plotted in the (r, x) space as depicted in Fig. 5.3. It can be seen that the expected profit is jointly concave in the prices r and x and forms a single peak at $r = r^*$ and $x = x^*$. For very low values of r and x (bottom left corner), the asset provider expects to earn a highly

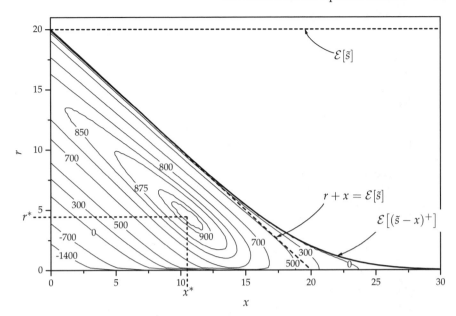

Fig. 5.4. Optimal pricing policy of the asset provider in the base case: Iso-profit lines of Π_A in the (r, x) space. The contour plot corresponds to Fig. 5.3. The optimal contract is an option contract ($r* > 0$, $x^* > 0$).

negative profit, which is not shown in Fig. 5.3. For high values of both r and x, the expected profits level off. This non-differentiability is due to the intermediary's participation constraint with regard to the contract market and can be analyzed more closely in Fig. 5.4 which represents a contour plot of Fig. 5.3.

As derived in Theorem 4.1, the participation of the intermediary in the contract market depends on the expected spot market premium. If the reservation fee is set greater than the expected spot market premium, i.e., $r > \mathcal{E}[(\tilde{s} - x)^+]$, then the intermediary will not reserve any capacity, and the asset provider is left with his spot market business only. The expected-profit function of the asset provider in (4.19) reduces to

$$\Pi_A = \mathcal{E}\left[(\tilde{s} - t)\tilde{M} - fK\right] \quad \text{for} \quad N = 0 \tag{5.5}$$

and results for the base case in $\Pi_A = -200$. In Fig. 5.4 this is the case for the area right and above the curve marked $\mathcal{E}[(\tilde{s} - x)^+]$. Since under this scenario the demand in the spot market is unlikely to be sufficient to cover the asset provider's fixed cost, he expects a negative profit.

It can also be seen in Fig. 5.4 that though the expected-profit function overall is rather steep, it is rather flat around its maximum such that small

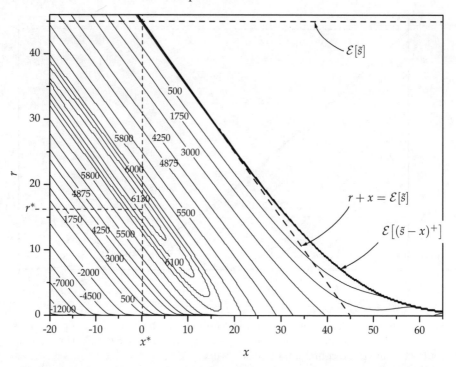

Fig. 5.5. Optimal pricing policy of the asset provider if the expected spot price is high ($\mathcal{E}[\tilde{s}] = 45$, $\sigma_{\tilde{s}} = 15$; all other variables as in the base case): Iso-profit lines of Π_A in the (r, x) space. The optimal contract is a fixed-commitment contract ($r^* > 0$, $x^* = 0$). The unconstrained solution to the optimization problem would be $r = 20.57$ and $x = -4.16$.

deviations from the optimal solution do not have a significant impact on profits. Since the optimal tariff has to be determined by numerically solving the system of equations provided in Theorem 4.2, this finding is of practical relevance for the required precision of the numerical solution algorithm.

One further observation is particularly interesting if the asset provider's objective is – instead of maximizing profits – obtaining a certain target profit and capacity utilization level. Given the concavity of the objective function, multiple combinations of r and x exist for the profit levels below the maximum profit that obtain the exactly same expected profit for the asset provider. For example, if it were the objective to obtain $\Pi_A \geq 900$ and holding x constant at $x = 10.6$, the asset provider could set $3.7 \leq r \leq 5.2$ and make use of this bandwidth in contract negotiation with the intermediary.

In Fig. 5.4, solving the asset provider's optimization problem gives an interior solution, i.e., the peak of the three-dimensional objective function falls within the quadrant for which $r > 0$ and $x > 0$ and the Kuhn-Tucker conditions in Theorem 4.2 result in $\ell_1, \ell_2 = 0$. By way of comparison, Fig. 5.5

shows an example for the case that $\ell_1 = 0$ and $\ell_2 > 0$. The parameters underlying Fig. 5.5 correspond to the base case with the exception of $\mathcal{E}[\tilde{s}] = 45$ and $\sigma_{\tilde{s}} = 15$ (see also Sec. 5.4.3.3). Here, the peak of the three-dimensional objective function falls outside the feasible region for x and the solution to the optimization problem is of the form $r > 0$ and $x = 0$ (more precisely, it is $r^* = 16.58$ in Fig. 5.5; see also Fig. 5.28) and the asset provider offers a fixed-commitment, not an option contract to the intermediary.

5.3 The Benefit of the Contract Market

Prior to discussing the two different types of capacity contracts (fixed-commitment and capacity-option contract), this section illustrates the benefit of engaging in advance sale of capacity altogether. To this end, the expected profits of buyer and seller when buying and selling only in the spot market (Scenario S) are compared with their respective performance when closing in addition a fixed-commitment contract (Scenario FS).

The comparison is conducted by means of a comparative static analysis (see also p. 80). The asset provider's capacity K is changed while leaving all other variables at the base-case values. Though in the model capacity is assumed to be fixed and can be altered beyond the time horizon of the model only, the analysis of different levels of capacity yields additional insight since changing capacity K means *ceteris paribus* changing the ratio of available capacity to total demand, i.e., $K/(\tilde{D}_C + \tilde{D}_S)$.

5.3.1 Optimal Pricing and Reservation Policies

The optimal pricing policy of the asset provider with regard to the fixed-commitment contract is indicated by the solid line marked w^* in Fig. 5.6. The dashed line marked N^* indicates the optimal reservation policy of the intermediary under this regime.

For low levels of capacity, demand in the spot market is (with expected demand being $\mathcal{E}[\tilde{D}_S] = 200$ in the base case) sufficient to fill capacity such that the optimal capacity price w^* is close to the expected spot price $\mathcal{E}[\tilde{s}]$. The intermediary has therefore little incentive to reserve capacity, resulting in low values for N^*. When the asset provider has more capacity, the optimal fixed-commitment price decreases considerably below the expected spot price, making the intermediary reserve more capacity. Thus N^* increases in K.[2] Furthermore, because of the price-responsiveness of contract market demand, contract market demand increases as w^* decreases. This is

---

[2] Considering the result for N^* obtained in (4.12), N^* is not directly a function of K. However, N^* is a function of the optimal tariff which is given implicitly by the set of equations in Theorem 4.2 and (5.4), respectively, where $\Phi(z_B)$ is a function of B

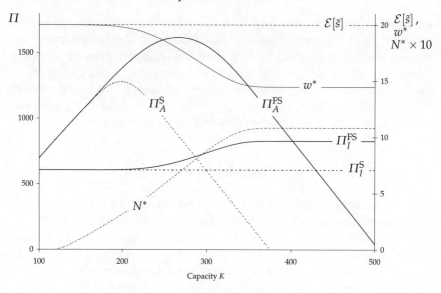

Fig. 5.6. Comparison of pure-spot market (S) and fixed-commitment contract scenario (FS) for varying levels of capacity K: If in addition to the spot market a fixed-commitment contract is closed, no party forfeits profit; for ample capacity, both parties are made better off.

due to the fact that lower values of w mean lower capacity procurement cost for the intermediary which he passes on to his end customers via a lower selling price p (see Sec. 4.2.3, where a fixed markup has been assumed).

For high levels of capacity, however, the optimal fixed-commitment price w^* declines to a level where the increase in volume induced by the price reduction no longer offsets the revenue loss associated with the price reduction. The optimal fixed-commitment price therefore levels off and so does the optimal number of reservations N^*.

5.3.2 Expected Profits

The profits both market participants can expect to earn in the pure spot-market scenario are depicted by the thick dashed lines in Fig. 5.6 marked Π_A^S for the asset provider and Π_I^S for the intermediary.

The expected profit of the intermediary is independent of the amount of capacity K offered by the asset provider because the intermediary can satisfy his demand from the entirety of all spot market sellers' capacity offerings. The expected profit of the asset provider is concave in the capacity

---

and B is a function of K. Thus, N^* is (indirectly) a function of K. However, given the form of (4.12) and the implicit formulation of r^*, x^*, and w^*, respectively, a differentiable analytical expression of this function cannot be found.

offered. For low levels of capacity, the asset provider can only partially satisfy the demand from the spot market and captures more of the spot market potential when the level of capacity is increased. For high levels of capacity, however, the spot market demand is insufficient to fill capacity K, leading to lower expected profits Π_A^S due to the fixed cost f associated with every unit of capacity.

For the fixed-commitment contract scenario, the expected profits are depicted by the thick solid lines in Fig. 5.6 marked Π_A^{FS} for the asset provider and Π_I^{FS} for the intermediary.

While the intermediary is gaining access to cheaper capacity in the contract market, his expected profit increases as compared to the pure spot-market scenario. This effect is amplified by demand stimulation through the decrease in w^* and, consequently, p and levels off when w^* and N^* converge. The shape of the asset provider's profit does not change when the asset provider adds fixed-commitment contracts to his capacity offerings. However, his expected profit increases for a wider range of capacity K as the asset providers taps an additional market – the contract market – and can generate additional sales.

Comparing the profits to be expected under the two scenarios shows that both the intermediary and the asset provider benefit for all levels of capacity from participating in the contract market. Thus the introduction of the fixed-commitment contract market in addition to the spot market is *Pareto improving*. The term *Pareto improvement* refers to a situation that makes everyone at least as well off and at least one party better off than in the initial situation (cf. Varian 1999, p.15).

5.3.3 Capacity Utilization

In fixed-capacity industries, one measure that is often used as a performance indicator is capacity utilization, although it can be shown that maximizing capacity utilization is not necessarily in line with profit maximization (see Sec. 5.4.1). Define expected capacity utilization κ as the relation of capacity expected to be used to total capacity, i.e.,

$$\kappa \equiv \frac{\mathcal{E}\left[(\tilde{E} + \tilde{M})\right]}{K}. \tag{5.6}$$

Fig. 5.7 plots expected capacity utilization κ over the amount of capacity offered. For both the pure-spot and the fixed-commitment contract scenario, capacity utilization decreases with increasing levels of capacity as one would expect from the definition in (5.6). Not surprisingly, the decline of κ^{FS} starts at higher capacity levels than of κ^S. The direct comparison between the two scenarios by $\Delta\kappa \equiv \kappa^{FS} - \kappa^S$ shows that the increase in capacity

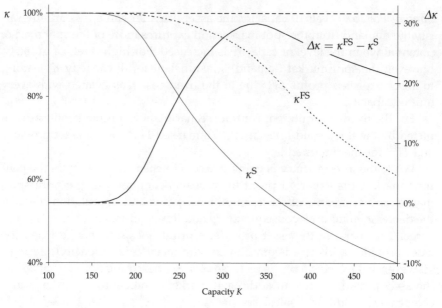

Fig. 5.7. Comparison of pure-spot market (S) and fixed-commitment contract scenario (FS) for varying levels of capacity K: Expected capacity utilization κ is higher if the asset provider offers capacity in spot and contract market. When more capacity is offered, utilization decreases.

utilization amounts here to up to 30%. This effect mainly results from the additional demand the asset provider encounters when he taps the contract market.[3]

5.4 Comparative Static Analysis of Contract Types

Having demonstrated the benefit of a combination of spot and fixed-commitment contract market over the pure spot-market scenario, this section extensively analyzes the two different types of long-term capacity contracts, namely fixed-commitment and capacity-option contract, with regard to the optimal type of contract and the determinants of the contract value.

The method chosen for this analysis is the comparative static analysis. By changing the underlying data of the model one at a time it can be shown in what way the model results depend on the underlying data. A formal mathematical analysis has to ruled out given the non-closed form of the

---

[3] In the following section, it is always assumed that the asset provider sells into the contract *and* spot market, i.e., similar effects resulting from the fact that the asset provider simply enlarges his customer base by tapping a new market do not occur.

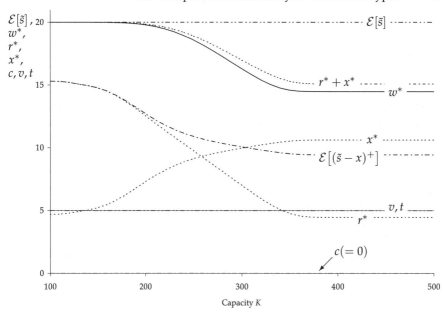

Fig. 5.8. Optimal pricing policy of the asset provider for varying capacity levels: The more capacity is available, the lower the reservation fee becomes.

optimal tariff in Theorem 4.2 that hinders a differentiation of the optimal reservation and execution fee with respect to other variables of the model.

Besides the parameters λ and f, all exogenous variables of the model are included in the following comparative static analysis. The intermediary's markup λ und and the fixed cost f are used for the calibration of the model only and do not qualitatively influence the structural results of the model. The basis of the analysis is the base case as specified in Table 5.1.

The analysis is structured into four parts: Firstly and in greater detail than the subsequent parts, the influence of the capacity size K on the two contract types is being looked at. Secondly, the variables shaping the contract market are considered, namely contract market size a, demand uncertainty as measured by $\sigma^2_{\tilde{D}_C}$, and price responsiveness b. Then, the analysis turns to the spot market with mean and variance of spot market demand ($\mu_{\tilde{D}_S}$, $\sigma^2_{\tilde{D}_S}$) and price ($\mu_{\tilde{s}}$, $\sigma^2_{\tilde{s}}$). Finally, it is shown how the cost structure (parameters c, t, and v) of the asset provider influences the model's results.

5.4.1 Capacity Size

Optimal Pricing Policy

Fig. 5.8 shows the optimal pricing policy under the fixed-commitment contract (solid line marked w^*) and under the option contract (dashed lines marked r^* and x^*). For better orientation, the diagram also displays the variable costs for short-term (spot sales, t) and long-term (sales under contract, v) capacity provision, the variable cost per accepted reservation c (assumed to be zero in the base case), and the expected spot price $\mathcal{E}[\tilde{s}]$.

The shape of the curve $r^* + x^*$ resembles the progression of the fixed-commitment price w^*, however, for high levels of capacity, the sum of the optimal reservation and execution fee exceeds the fixed-commitment price. One can further observe that the scarcer capacity is, the higher the optimal reservation fee; and, vice versa, the more capacity is available, the lower the optimal reservation fee. The economic explanation for this is that low levels of capacity imply a high ratio of aggregated demand $\tilde{D}_S + \tilde{D}_C$ to available capacity K. The asset provider expects sufficient demand, i.e., enough buyers that are willing to pay the spot price or a contract price close to the spot price to fully load capacity. The optimal option contract for scarce capacity therefore features a high upfront component, i.e., reservation fee. More specifically, the asset provider sets the reservation fee as high as he can to still assure the participation of the intermediary in the contract market. In Fig. 5.8, r^* is thus set close to the expected spot market premium $\mathcal{E}[(\tilde{s} - x)^+]$ for low levels of capacity.

The combination of high reservation and low execution fee can be interpreted as an indication that market power is distributed in favor of the seller. Given the trade-off between reservation and execution fee, the buyer, if he were to choose, would prefer the combination of a low reservation fee and a high execution fee since (i) a high execution fee means a greater chance of the spot price falling below the execution fee and thus a greater chance of lower capacity procurement cost and (ii) since a low reservation fee means low sunk cost in the case of non-execution. The preferences of the seller are contrary: A high reservation fee means a lower exposure to contract-market-demand and spot-price risk, while a low execution fee reduces the risk of options not being executed.

Market power shifts to the buyer when capacity is rather ample, as it is the case for the right hand side of Fig. 5.8, than scarce. At approximately $K = 300$, capacity equals the sum of expected demands in contract and spot markets, i.e., $K \approx \mathcal{E}[\tilde{D}_C] + \mathcal{E}[\tilde{D}_S]$ (see also Fig. 5.11). The asset provider then has to concede to the intermediary a lower reservation fee and collects his margin by the higher execution fee and therefore bears more risk than with scarce capacity.

Optimal Reservation Policy

The optimal reservation policy of the intermediary is shown in Fig. 5.9 by the lines marked N^*. The progression of the optimal number of reservations over varying levels of capacity is similar under the two scenarios, with the

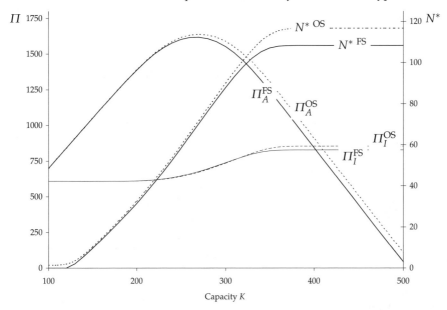

Fig. 5.9. Optimal reservation policy and expected profits for varying capacity levels: The optimal number of reserved capacity units N^* is always greater for the option contract scenario. For medium levels of capacity, the option contract is slightly worse for the intermediary, for ample capacity better than the fixed-commitment contract. The asset providers benefits for medium and high capacity levels.

optimal number of reservations under the option contract scenario being strictly greater than under the fixed-commitment contract scenario. However, for high levels of capacity, the number of options purchased exceeds the number of fixed-commitment purchases more than for low levels of capacity. This effect is a direct consequence from the changing ratio of reservation and execution fee.

As derived in Sec. 4.3, the number of reservations made by the intermediary is the result of trading off underage and overage cost. The optimal service level $F_C(N^*)$ of the intermediary, i.e., the probability that the reserved capacity N^* accommodates the entire demand from end customers, has been found to be given by $F_C(N^*) = c_u/(c_o + c_u)$. For the fixed-commitment contract scenario, the underage cost is $c_u^{FS} = \mathcal{E}[\tilde{s}] - w^*$. The overage cost is given by the price per reservation, i.e., $c_o^{FS} = w^*$, and the service level thus is $F_C(N^{* FS}) = 1 - w^*/\mathcal{E}[\tilde{s}]$. Since w^* is decreasing in K, the service level of the intermediary is increasing and so is $N^{* FS}$ (see Fig. 5.10). The price sensitivity of demand $\tilde{D}_C$ further adds to this effect.

Even though the total cost to the intermediary for one unit of capacity bought and used under the option contract is higher than under the fixed-commitment contract ($r^* + x^* \geq w^*$), the intermediary purchases more ca-

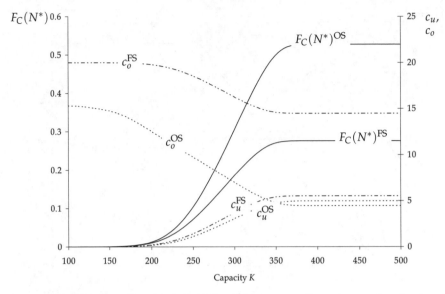

Fig. 5.10. Optimal service level $F_C(N^*)$ of the intermediary for varying levels of capacity: In the option contract scenario, the optimal service level is higher than in the fixed-commitment contract scenario.

pacity options than he reserves capacity under the fixed-commitment contract ($N^{*\,OS} > N^{*\,FS}$) because the optimal service level of the intermediary is higher under the option contract scenario. Under this scenario, the reservation fee constitutes the overage cost ($c_o = r^*$) and the difference between the expected spot market premium and the reservation fee represents the underage cost ($c_u = \mathcal{E}[(\tilde{s} - x)^+] - r$); thus $F_C(N^{*\,OS}) = 1 - r/\mathcal{E}[(\tilde{s} - x)^+]$. Consequently, if as in Fig. 5.8 the reservation decreases more than the execution fee increases, the service level $F_C(N^{*\,OS})$ increases and so does the optimal number of reservations $N^{*\,OS}$.

Expected Profits and Welfare Analysis

Figure 5.9 compares the expected profits earned by the two parties when following their respective optimal policy under the option and fixed-commitment contract scenario. The shapes of the profit functions with option contracts correspond to the shapes with fixed-commitment contract described in Sec. 5.3.2. For medium and high capacity levels, the asset provider expects to earn higher profits when using option contracts, for ample levels of capacity the intermediary, too.

Given the concavity of the asset provider's expected profits in capacity size K, it would be possible to numerically determine the optimal capacity size. However, as discussed in Sec. 2.1.2, capacity cannot easily be adjusted and has therefore been assumed to be fixed within the time horizon of the

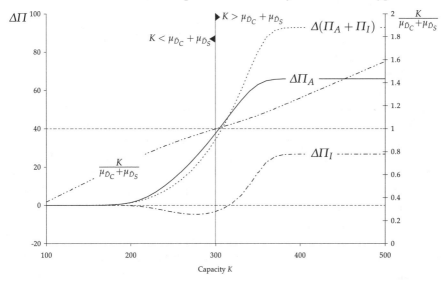

Fig. 5.11. Comparison of expected profit differentials for varying levels of capacity: Total welfare improvement increases with the amount of capacity offered by the asset provider. For medium levels of capacity, the intermediary's welfare dips slightly and then increases when capacity is ample.

model (see Chap. 4) and is thus not modeled as a decision variable. Furthermore, capacity can only be changed in large increments and may be determined by external factors, e.g., demand for passenger transport (see likewise Sec. 2.1.2), or market-share and strategic considerations. It is hence possible that capacity in reality deviates from the asset provider's profit maximum in Fig. 5.9.

For a closer analysis of the impact of the choice of contract type on profits for varying capacity sizes, Fig. 5.11 plots the differences in expected profits for the two market participants as well as the differences of their joint expected profit. To this end, let $\Delta\Pi_i \equiv \Pi_i^{OS} - \Pi_i^{FS}$ with $i \in \{A, I\}$. For all levels of capacity, the asset provider expects a non-negative profit differential $\Delta\Pi_A$. While the absolute advantage of option contracts levels off for high capacity levels, the relative advantage of option contracts is the higher, the more capacity the asset provider offers since absolute expected profits Π_A decrease for high level of capacity. For $K = 400$, e.g., the asset provider can increase expected profits by 8% through using option contracts, for $K = 450$ even by 15%. For ample levels of capacity, the asset provider thus will strongly prefer option contracts over fixed-commitment contracts and is indifferent between the two contract forms for scarce capacity only since, in the latter case, he can most likely sell off his capacity on the spot market at a higher margin.

If capacity is ample, the intermediary, too, expects higher profits when purchasing capacity options instead of signing a fixed-commitment contract. However, for medium levels of capacity – when the sum of market demands is in the order of magnitude of capacity ($K \approx \mu_{\tilde{D}_C} + \mu_{\tilde{D}_S}$, see Fig. 5.11) – the expected profit differential $\Delta \Pi_I$ is slightly negative. Thus, the intermediary will per se not have an incentive to sign an option contract. The decrease of expected profits amounts to up to 0.7%. However, since the joint profit differential $\Delta(\Pi_A + \Pi_I)$ is positive for all capacity levels, it would be possible to achieve a Pareto improvement, e.g., by a side payment, if the two parties were to negotiate the contract. By this means, the asset provider could compensate the intermediary for the financial disadvantage of the option contract and thus overcome resistance against the contract and tariff that in total achieves higher profits.

To conclude, total welfare is increased over all capacity levels. However, not for all levels of capacity a Pareto improvement is achieved in the first place since the intermediary forfeits some of his profits when instead of fixed-commitment contracts option contracts are offered at medium levels of capacity. However, by a mutually negotiated agreement, a Pareto improving situation could be achieved.

Capacity Utilization

As already seen in Sec. 5.3.3, capacity utilization κ decreases when more capacity is offered. Fig. 5.12 illustrates that this holds also true for the option contract scenario. Comparing capacity utilization under the two contract scenarios yields the curve marked $\Delta\kappa$ which is defined as $\Delta\kappa \equiv \kappa^{OS} - \kappa^{FS}$. It can be seen that capacity utilization using option contracts is slightly below capacity utilization using fixed-commitment contracts, reflecting the chance that options are not executed by the intermediary due to a favorable spot-price development or an unfavorable demand development. Under this comparative static analysis with varying levels of capacity, the drop in expected capacity utilization is only marginal.[4] Nevertheless, option contracts contradict the common but naïve understanding that keeping capacity utilization high must be in line with maximizing profits.

5.4.2 Contract Market

Having studied how variations of the capacity offering K impact the pricing of fixed-commitment and capacity-option contracts and the relative advantage of one over the other, the comparative static analysis continues in this section by studying how the characteristics of the contract market influences

---

[4] For other parameter constellations, expected capacity utilization can also be higher for a capacity-option contract than for fixed-commitment contract.

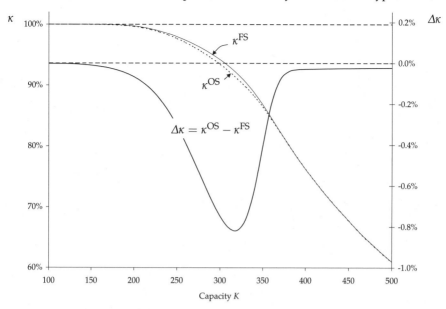

Fig. 5.12. Capacity utilization κ for varying capacity levels: For medium levels of capacity, capacity utilization is slightly lower if the asset provider sells option rather than fixed-commitment contracts. When more capacity is offered, utilization generally decreases.

the choice and pricing of the two types of capacity agreements. Three characteristics of the contract market are of foremost importance: the market size[5], the uncertainty associated with demand in the contract market, and the price responsiveness of demand.

5.4.2.1 Size of the Contract Market

As introduced in Sec. 4.2.2, the size of the contract market is given by the ordinate intercept a of the contract market demand function. More precisely, a gives the expected theoretic maximum size of the contract market if p were equal to 0. The actual size of the contract market will thus in expectation always be smaller than a if $p > 0$. Nevertheless, for any given $p \geq 0$ it holds that expected contract market demand increases in a.

Optimal Pricing and Reservation Policies

For the fixed-commitment contract, Fig. 5.13 shows that capacity reservations become the more expensive the greater the contract market is (w^* is increasing in a) because demand increases while supply remains unchanged.

---

[5] Market size is in the following determined by the demand side of the contract market, not by the supply side.

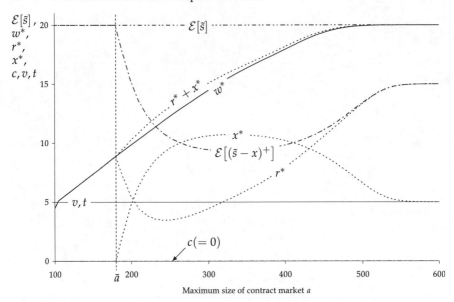

Fig. 5.13. Optimal pricing policy for varying contract market sizes: If the maximum size of the contract market is smaller than threshold size $\bar{a}$, the asset provider does not offer option contracts. For high values of a, the execution fee x^* approaches variable cost v.

For large sizes of the contract market, the optimal capacity price w^* approaches the expected spot price $\mathcal{E}[\tilde{s}]$ but never exceeds it since then the intermediary would turn for his entire capacity procurement to the spot market due to lower expected procurement cost.

The total price $r^* + x^*$ under the capacity-option contract is for all levels of a at least as high as the price under the fixed-commitment contract. As already discussed in Sec. 5.4.1, $r^* + x^*$ may exceed w^* without making the intermediary worse off since in expectation not all options are exercised.[6] If the maximum contract market size is smaller than the threshold size $\bar{a}$, Fig. 5.13 shows that $r^* = w^*$ and $x^* = 0$. This means that the asset provider does not offer option contracts but fixed-commitment contracts only if the contract market is smaller than $\bar{a}$ because the constraint $x \geq 0$ in (4.19) becomes binding.

Lemma 5.1. *The threshold value $\bar{a}$ is implicitly given by the solution to the following set of equations for $(r^*, \bar{a})$ with $x = 0$:*

---

[6] w^* can be shown to be exceeded even by the total expected price of an executed option, i.e., $r^* + \mathcal{E}[\min(\tilde{s}, x)]$. This is due to the fact that non-execution does not only occur if $\tilde{s} < x$ (no execution at all), but also if $\tilde{D}_C < N$ (partial execution).

$$\bar{a} - b(r^* + \lambda) + z_N \sigma_{\tilde{D}_C} - \left[r^* - c - (\mu_{\tilde{s}} - t)(1 - \Phi(z_B)) \right] \left[\frac{\sigma_{\tilde{D}_C}}{\varphi(z_N)\mu_{\tilde{s}}} + b \right]$$
$$+ v \left[\frac{\sigma_{\tilde{D}_C}(1 - \Phi(z_N))}{\varphi(z_N)\mu_{\tilde{s}}} + b \right] = 0 \quad (5.7)$$

$$\bar{a} - b(r^* + \lambda) - \sigma_{\tilde{D}_C} L(z_N)$$
$$- \left[r^* - c - (\mu_{\tilde{s}} - t)(1 - \Phi(z_B)) \right] \left[\frac{\sigma_{\tilde{D}_C}(1 - \Phi(z_N))}{\varphi(z_N)\mu_{\tilde{s}}} + b \right]$$
$$+ v \left[\frac{\sigma_{\tilde{D}_C}(1 - \Phi(z_N))^2}{\varphi(z_N)\mu_{\tilde{s}}} + b \right] = 0 \quad (5.8)$$

Proof. At $\bar{a}$, it is $r = r^*$ and $x = x^* = 0$ and thus $\partial \Pi_A / \partial r = \partial \Pi_A / \partial x = 0$ (first order condition, unconstrained optimization). Let $a = \bar{a}$. Taking $\partial \Pi_A / \partial r$ from (4.30) with (4.31)–(4.33) and setting $\partial \Pi_A / \partial r = 0$ and $x = 0$ gives (5.7). Taking $\partial \Pi_A / \partial x$ from (4.34) with (4.35)–(4.36) and setting $\partial \Pi_A / \partial x = 0$ and $x = 0$ gives (5.8). □

For contract market sizes greater than $\bar{a}$, the asset provider does offer option contracts. The optimal reservation fee r^* first decreases in a, then increases, and finally levels off. This behavior can be explained by considering the complementary nature of the optimal execution fee x^*. At first, r^* decreases to boost the number of reservations sold. This implies an increase in x^* to make up for the margin lost by the decrease of r^*. However, an increasing x^* means an increasing chance of the spot price being below the execution fee and thus of non-execution of options. With the contract market becoming larger, the asset provider can raise the price of capacity by increasing r^*, again due to the fact that with increasing contract market size demand becomes larger with supply being fixed. The increase of r^* comes to a halt when r^* approaches the expected spot market premium $\mathcal{E}[(\tilde{s} - x)^+]$ and when x^* decreases to such an extent that it approaches the variable cost v caused by option execution. Hence, for large values of a, the asset provider collects his entire margin from the reservation fee only and charges an execution fee that equals his marginal cost.

Expected Profits and Welfare Analysis

Fig. 5.14 shows the optimal reservation policy of the intermediary and the profits expected to be earned by the market participants for varying levels of a. The capacity quantities N^* reserved by the intermediary generally increase in a: the greater the market, the higher the expected demand, the

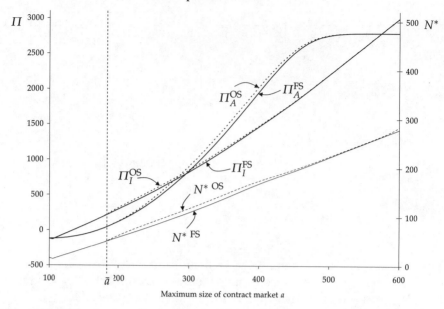

Fig. 5.14. Optimal reservation policy and expected profits for varying contract market sizes: A greater contract market entails higher expected profits and higher quantities of reserved capacity.

more need for capacity reservation. For market sizes greater than $\bar{a}$, the number of capacity options exceeds the number of reservations under the fixed-commitment contract. This is again due to a favorable shift of the relation of underage and overage cost faced by the intermediary.

The expected profits of the intermediary in Fig. 5.14 generally increase in a since greater values of a mean higher demand. For the same reason, the asset provider's expected profits are generally increasing in a. However, for high values of a, the lines marked $\mathcal{E}[\Pi_A]$ level off because, in combination with his spot market sales, the asset provider reaches his capacity limit such that additional demand from the contract market no longer increases his expected profits.

Fig. 5.15 compares the expected profit increments between the fixed-commitment and the option-contract scenario. The profit differential for both market participants is always non-negative. For high values of a, the profit differential is negligibly small. However, for contract market sizes between $\bar{a}$ and values of a for which the price for reserved capacity approaches the expected spot price, option contracts improve the expected profits of both option buyer and seller. The maximum relative profit improvement potential of option contracts compared to fixed-commitment contracts amounts to 8–9% in the analyzed base case for both the asset provider and intermediary.

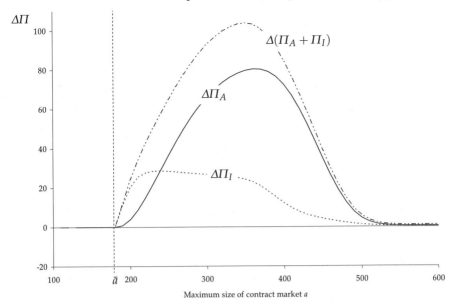

Fig. 5.15. Welfare analysis for varying contract market sizes: For all values of a, both market participants are at least as good off with a capacity-option contract as with a fixed-commitment contract. For medium contract market sizes, the Pareto improvement potential is highest.

5.4.2.2 Demand Uncertainty in the Contract Market

Demand uncertainty in the contract market is measured by contract market demand variance $\sigma^2_{D_C}$ which is identical with $\sigma^2_{\tilde{\epsilon}}$ (see (4.2)).

Optimal Pricing and Reservation Policies

Fig. 5.16 depicts the optimal pricing policy of the asset provider. The higher the demand uncertainty, the lower the capacity price w^* under the fixed-commitment contract. Under the option contract, the reservation fee r^* decreases in the demand variance, the execution fee increases x^*. The optimal policy thus entails that the asset provider takes on some of the demand risk. This becomes particularly obvious in the case of the option contract where more and more of the total price charged by the asset provider shifts from the upfront price component (the reservation fee) to the component which is not received for sure (execution fee). To compensate for the non-execution of options that becomes more likely with increasing demand variance, the price premium $(r^* + x^*) - w^*$ of capacity used on the basis of an option contract as compared to a fixed-commitment contract becomes larger.

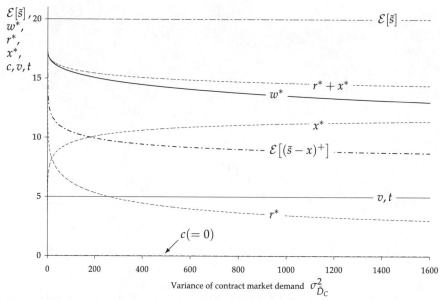

Fig. 5.16. Optimal pricing policy for varying levels of contract market demand variance: As demand variance increases, the asset provider raises the execution fee x^* in favor of a low reservation fee r^*.

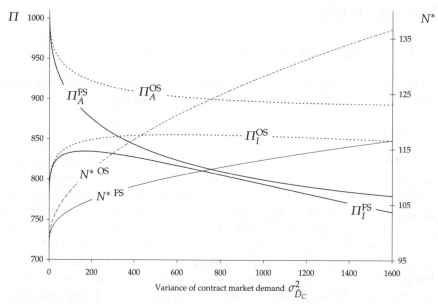

Fig. 5.17. Optimal reservation policy and expected profits for varying levels of contract market demand variance: While expected profits under the fixed-commitment contract decline for high values of demand variance, expected profits under the option contract stabilize at a significantly higher level.

Expected Profits and Welfare Analysis

The effect of the pricing policy on the optimal reservation policy of the intermediary can be inferred from Fig. 5.17. For both contract types, the optimal amount of reserved capacity N^* increases as demand variance increases. Reverting to the calculation of $N^* = \mu_{\tilde{D}_C} + z_N \sigma_{\tilde{D}_C}$ from (4.16) shows that N^* must clearly increase when variance $\sigma_{\tilde{D}_C}$ increases. However, Fig. 5.17 shows that, in case of the capacity-option contract, it does so at a higher rate. This can be explained by the evolution of underage and overage cost when demand variance increases. The underage cost c_u (see p. 62) increases slowly in demand variance and amounts to a similar value for both types of contract. Also for both types of contract, the overage cost c_o decreases in demand variance, however, for the option contract it amounts to a significantly lower value. Hence the service levels (and thus z_{N^*}; see p. 62) increase in demand variance. Due to the significantly lower overage cost, it does so more strongly in the case of the capacity-option contract.

Fig. 5.17 also depicts the market participants' expected profits over varying levels of demand variance. As demand variance increases from 0, the intermediary's expected profits initially increase due to the decrease in w^* and $r^* + x^*$, respectively. As demand variance increases further, the intermediary's expected profits decline in demand variance as do the expected profits of the asset provider in general. However, the decline in profits is for both parties much stronger in the case of the fixed-commitment contract. For the option contract, profits are much more stable even for high levels of demand variance. Thus option contracts hedge both seller and buyer against variance of contract market demand. This is achieved by the effective risk sharing rendered possible by the split tariff of the option contract.

To compare the expected profits among the fixed-commitment and option contract scenario, Fig. 5.18 plots the differences in expected profits for the two market participants as well as the difference in joint expected profits. In line with the above observations, Fig. 5.18 shows that the differences in expected profits are increasing in demand variance. The value of capacity options thus correlates positively with contract market demand uncertainty. For $\sigma_{\tilde{D}_C}^2 = 1600$, the relative increase of expected profits amounts to 15% for the asset provider and 12% for the intermediary.

This finding distinguishes capacity options from the nature of financial options (see Sec. 3.2.3.2). The value of financial (call) options increases in uncertainty only from the buyer's perspective, but decreases from the seller's point of view. Here, the value of capacity options increases for *both* parties in a particular source of uncertainty.

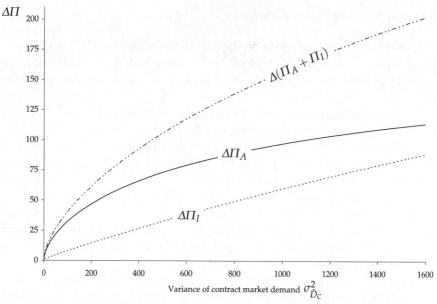

Fig. 5.18. Welfare analysis for varying levels of contract market demand variance: The incremental expected profits and thus the value of capacity options increase in the demand variance.

5.4.2.3 Price Responsiveness of Contract Market Demand

The price responsiveness of contract market demand is given by the slope b of the contract market demand function (cf. Varian 1999, p. 265). High values of b mean a steeper demand function and thus that demand from end customers decreases strongly when the price increases. However, taking b as the only measure for price responsiveness is problematic since its values depend on the units demand and price are measured in. For this reason, in microeconomics often the concept of price elasticity is used since it represents a unit-less measure of price responsiveness and can thus be used for comparison among different products which are measured in different units (cf. Hyman 1993, p. 137). The price elasticity η of contract market demand is defined as the ratio of a relative change in contract market demand to a relative change in price:

$$\eta \equiv \frac{\Delta \tilde{D}_C / \tilde{D}_C}{\Delta p / p}. \tag{5.9}$$

According to this definition, the price elasticity is different at different points on the demand function as well as for different values of b. "High" and "low" values of price elasticity refer in the following always to the *absolute* value of η, i.e., $|\eta|$. For $|\eta| < 1$, contract market demand is said

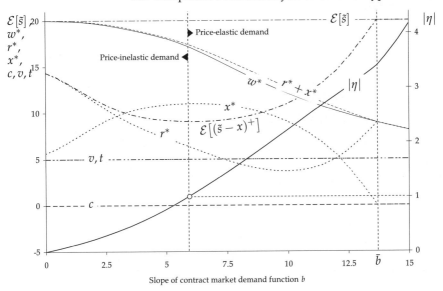

Fig. 5.19. Optimal pricing policy for varying responsiveness of contract market demand, measured by slope b of the contract market demand function: If $b > \bar{b}$, the asset provider does not offer option contracts.

to be (price-) *inelastic* since a 1% increase in price leads to a reduction of demand by less than 1%. For $|\eta| \geq 1$, contract market demand is called (price-) *elastic* since demand decreases more than proportionally when price increases.

For the linear demand function assumed in Sec. 4.2.2 ((4.1)), the (point) price elasticity of mean demand is given by:

$$\eta = \frac{\partial \mu_{\bar{D}_C}}{\partial p} \frac{p}{\mu_{\bar{D}_C}} = \frac{-bp}{a - bp}. \tag{5.10}$$

Optimal Pricing and Reservation Policies

Fig. 5.19 shows the optimal pricing policy of the asset provider as a function of the price responsiveness of contract market demand, measured by the slope b of the contract market demand function. To make this measure more meaningful, Fig. 5.19 also indicates the price elasticity η on the secondary ordinate. If demand is not responsive to price, i.e., $b \to 0$, the asset provider charges a price w^* in the limit equal to the expected spot price $\mathcal{E}[\bar{s}]$. As b increases, it is optimal for the asset provider to lower the price w in order to compensate for the decline in volume that goes along with increasing price responsiveness of demand. The course of the optimal total capacity price $r^* + x^*$ under the option contract is similar to w^*, with $r^* + x^* > w^*$ up to threshold level $\bar{b}$. If price responsiveness is greater than $\bar{b}$, the asset provider no longer offers option contracts. Like in the analysis for varying sizes of the contract market (see Sec. 5.4.2.1), this is due to the fact that the constraint $x \geq 0$ takes effect.

Lemma 5.2. *The threshold value $\bar{b}$ is implicitly given by the solution to the following set of equations for $(r^*, \bar{b})$ with $x = 0$:*

$$a - \bar{b}(r^* + \lambda) + z_N \sigma_{\tilde{D}_C} - \left[r^* - c - (\mu_{\tilde{s}} - t)(1 - \Phi(z_B)) \right] \left[\frac{\sigma_{\tilde{D}_C}}{\varphi(z_N)\mu_{\tilde{s}}} + b \right]$$

$$+ v \left[\frac{\sigma_{\tilde{D}_C}(1 - \Phi(z_N))}{\varphi(z_N)\mu_{\tilde{s}}} + b \right] = 0 \quad (5.11)$$

$$a - \bar{b}(r^* + \lambda) - \sigma_{\tilde{D}_C} L(z_N)$$

$$- \left[r^* - c - (\mu_{\tilde{s}} - t)(1 - \Phi(z_B)) \right] \left[\frac{\sigma_{\tilde{D}_C}(1 - \Phi(z_N))}{\varphi(z_N)\mu_{\tilde{s}}} + b \right]$$

$$+ v \left[\frac{\sigma_{\tilde{D}_C}(1 - \Phi(z_N))^2}{\varphi(z_N)\mu_{\tilde{s}}} + b \right] = 0 \quad (5.12)$$

Proof. At $\bar{b}$, it is $r = r^*$ and $x = x^* = 0$ and thus $\partial \Pi_A / \partial r = \partial \Pi_A / \partial x = 0$ (first order condition, unconstrained optimization). Let $b = \bar{b}$. Taking $\partial \Pi_A / \partial r$ from (4.30) with (4.31)–(4.33) and setting $\partial \Pi_A / \partial r = 0$ and $x = 0$ gives (5.11). Taking $\partial \Pi_A / \partial x$ from (4.34) with (4.35)–(4.36) and setting $\partial \Pi_A / \partial x = 0$ and $x = 0$ gives (5.12). □

Up to slope $\bar{b}$, the asset provider does offer option contracts.[7] As b increases from 0, the optimal reservation fee r^* decreases from its maximum possible level (the expected spot market premium $\mathcal{E}[(\tilde{s} - x)^+]$, as established in Theorem 4.1) to stimulate sales of capacity reservations, while x^* increases to partially compensate for the decline in r^*. However, as b increases further, the increase of x^* slows down until x^* finally decreases,

---

[7] For sound decision making on the type of contract being offered and the respective pricing policy, the asset provider needs to collect information to draw conclusions about the demand function and, by this way, about price responsiveness. A common method of determining buyers' trade-off between competing products, suppliers, and product attributes (including price) nowadays is the *conjoint analysis* (cf. Green et al. 2001), dating back to the seminal research by Luce and Tukey (1964). See Brzoska (2003) for a recent and comprehensive treatise of the subject. Schmelter (2002) reports on the realization of a conjoint study for air cargo products at Lufthansa Cargo AG.

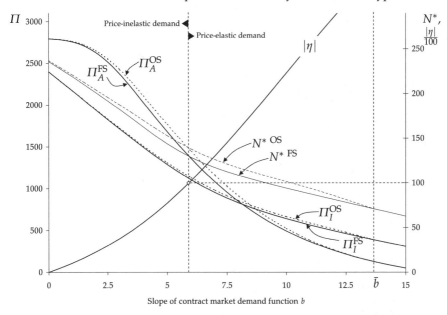

Fig. 5.20. Optimal reservation policy and expected profits for varying price respon-siveness of contract market demand, measured by slope b of the contract market demand function: The steeper the demand function, the lower the profits of both parties.

which can be explained by two reasons: First of all, the chance of x being greater than the realization of $\tilde{s}$ increases with x^* and thus the risk of non-execution of options; secondly, demand falls in x^*.

Expected Profits and Welfare Analysis

Fig. 5.20 gives insight into the intermediary's optimal reservation policy and the profits both parties can expect to earn for different levels of price respon-siveness of contract market demand. The higher the price responsiveness, the lower is the optimal number of reserved capacity units N^*. The decline in prices w^* and $r^* + x^*$, respectively, thus does not entirely offset the de-cline in demand caused by the increasing price responsiveness of demand. As in the preceding analyses, the optimal number of options $N^{* \, OS}$ is (up to $\bar{b}$) higher than the number of fixed-commitment reservations $N^{* \, FS}$ due to the higher service level resulting from significantly lower overage cost in the case of the option contract.

In Fig. 5.20, the expected profits of both parties generally decline in b. This is a consequence of the trade-off between sales volume and selling price. Since both decline in response to demand becoming more responsive to price, the asset provider's expected profits decrease. As it is the demand

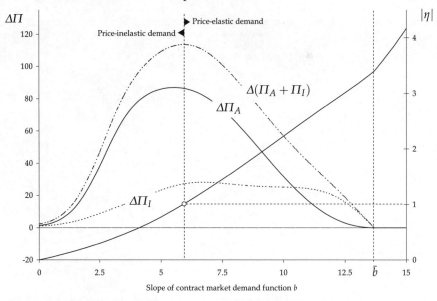

Fig. 5.21. Welfare analysis for varying price responsiveness of contract market demand, measured by slope b of the contract market demand function: Up to $b = \bar{b}$, both parties are better off when using capacity options instead of fixed-commitment contracts.

from the intermediary's customers whose price responsiveness is modeled by b, the decline of the intermediary's profit follows directly from the decline in revenue the intermediary experiences when demand becomes more responsive to price.

Fig. 5.21 shows the additional profits both market participants can expect when using capacity options instead of fixed-commitment contracts. Up to $b = \bar{b}$, both buyer and seller benefit from option contracts. For the base case, option contracts are Pareto improving. The relative increase of expected profit amounts up to 4% for the intermediary, for the asset provider up to 8%.

The fact that the highest joint performance improvement in Fig. 5.21 is around $|\eta| = 1$ is incidental for the base case and does not necessarily hold for other parameter constellations.

5.4.3 Spot Market

Having analyzed the influence of the characteristics of the contract market on the pricing and performance of long-term capacity contracts, the analysis now turns to the spot market. For the intermediary, the spot market represents an alternative way to procure capacity that he uses when the demand he faces from his customers exceeds the amount of capacity he has reserved with the asset provider or when the spot price is below the execution fee. For the asset provider, the spot market is an additional market to sell capacity that has not been sold in the contract market. How much capacity is kept for spot market sales depends on the attractiveness of the spot market, which is determined by

1. the size of the spot market, i.e., the expected demand in the spot market, and the uncertainty thereof;
2. the price earned in the spot market, i.e., the expected spot price and the uncertainty thereof.

The amount of capacity that is left over for the spot market depends on the amount of capacity sold in the contract market.[8] The asset provider controls this amount of pre-sold capacity by setting the tariff in the contract market accordingly.

5.4.3.1 Expected Spot Market Demand

To analyze the impact of the spot market size, the expected spot market demand $\mu_{\tilde{D}_S}$ is varied. Since the relative uncertainty of the spot market demand is determined by the coefficient of variation $\vartheta_{\tilde{D}_S} = \sigma_{\tilde{D}_S}/\mu_{\tilde{D}_S}$, varying $\mu_{\tilde{D}_S}$ implies changing the degree of uncertainty. However, if the analysis were conducted this way, the results would be ambiguous with regard to the cause of the effects observed since these could be attributable to the change in the size of the spot market or to the different degree of uncertainty (cf. Lariviere and Porteus 2001). To avoid this ambiguity, the coefficient of variation is, for the following analysis, kept constant at $\vartheta_{\tilde{D}_S} = \sigma_{\tilde{D}_S}/\mu_{\tilde{D}_S} = 1/3$ by changing $\sigma_{\tilde{D}_S}$ when changing $\mu_{\tilde{D}_S}$.[9]

Optimal Pricing and Reservation Policies

The curves marked w^* and $r^* + x^*$ in Fig. 5.22 show that if the spot market is small, the asset provider prices the capacity contracts lower than if the expected spot demand is high. As the expected spot demand increases, the total capacity prices w^* and $r^* + x^*$ converge the closer to the expected spot price $\mathcal{E}[\tilde{s}]$, the more the expected spot demand exceeds the asset provider's capacity K.

---

[8] For the time being, it is assumed that only the non-reserved share of total capacity is sold in the spot market, i.e., $B = K - N$. In Sec. 6.3, overbooking is introduced, allowing for $B \geq K - N$.

[9] Keeping $\sigma_{\tilde{D}_S}$ constant and thus letting $\vartheta_{\tilde{D}_S}$ decrease as $\mu_{\tilde{D}_S}$ increases does not change the general shape of the curves displayed in Figures 5.22 to 5.24 but rather compresses them horizontally. The general results derived in this section are thus independent of the choice of keeping either $\vartheta_{\tilde{D}_S}$ or $\sigma_{\tilde{D}_S}$ constant.

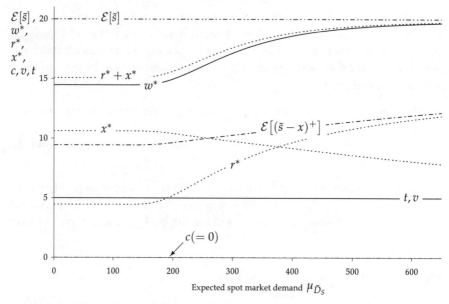

Fig. 5.22. Optimal pricing policy as a function of the expected demand in the spot market with constant coefficient of variation $\vartheta_{\tilde{D}_S} = 1/3$: Option contracts are being offered for all levels of expected spot market demand.

However, the asset provider participates in the contract market for all levels of $\mu_{\tilde{D}_S}$, even for very high values beyond the range shown in Fig. 5.22. This is because of the uncertainty in spot market demand as expressed by $\vartheta_{\tilde{D}_S} = 1/3$, which always leaves a non-negligible probability of demand $\tilde{D}_S$ being smaller than capacity K.

The higher the expected demand in the contract market, the less the asset provider depends on the contract market. Therefore, $r^* + x^*$ increases and the split tariff shifts more and more to the upfront component r^* while the – from the asset provider's perspective – not surely received component x^* becomes smaller. For high values of $\mu_{\tilde{D}_S}$, r^* and x^* converge against levels which depend on the value of $\vartheta_{\tilde{D}_S}$.

Fig. 5.23 displays the optimal reservation policy and expected profits resulting from this pricing policy. The optimal amounts of reserved capacity N^* decrease in $\mu_{\tilde{D}_S}$ as soon as the prices w^* and $r^* + x^*$ increase. Since the reservation fee r^* is always smaller than the fixed-commitment price w^*, the overage cost under the option contract is lower than under the fixed-commitment contract and induces a higher optimal service level. $N^{*\,OS}$ is thus greater than $N^{*\,FS}$ for all levels of $\mu_{\tilde{D}_S}$.

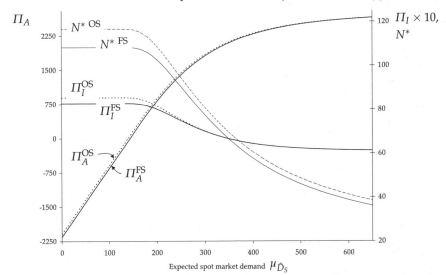

Fig. 5.23. Optimal reservation policy and expected profits as a function of the expected demand in the spot market with constant coefficient of variation $\vartheta_{\tilde{D}_S} = 1/3$: The higher the expected spot market demand, the more the asset provider retreats from the contract market and earns higher profits in the spot market.

Expected Profits and Welfare Analysis

The expected profits of the two market participants in Fig. 5.23 are affected quite differently by the expected size of the spot market. The expected profits of the asset provider depend strongly on the expected spot market demand. If expected demand is low, the asset provider suffers a severe loss due to under-utilization of his capacity. The higher the expected spot market demand, the higher the chance of fully utilizing capacity and thus the higher the expected profit.

The intermediary's expected profit is much less, but adversely affected by the expected spot market demand (note that, in Fig. 5.23, Π_I is plotted against the secondary ordinate to better show the curves' shape; plotting Π_I against the primary ordinate results in even flatter, almost horizontal curves). The intermediary's expected profit is not impacted by the demand volume in the spot market as such, but only indirectly via the asset provider's pricing policy that reacts to the spot market size. Since the price charged by the asset provider when reserving capacity with him increases in $\mu_{\tilde{D}_S}$, the capacity procurement cost of the intermediary increases and his profits decrease.

Fig. 5.24 compares differences in expected profits between option contracts and fixed-commitment contracts. For all sizes of the spot market considered here, option contracts are Pareto improving. The lower the expected

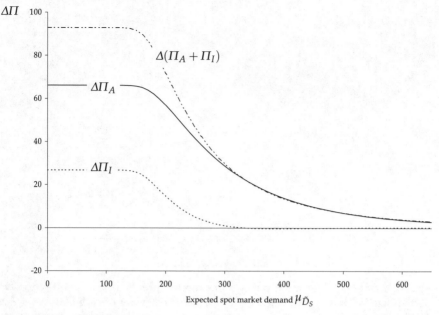

Fig. 5.24. Welfare comparison for varying expected levels of spot market demand with constant coefficient of variation $\vartheta_{\tilde{D}_S} = 1/3$: For all levels of expected spot demand, option contracts are Pareto improving.

spot market demand, the relatively better is the use of option contracts for both market parties. The higher the expected spot market demand, the less capacity is reserved anyway and the lower consequently the positive impact resulting from the choice of contract type. For low values of $\mu_{\tilde{D}_S}$, the relative improvement of expected profits amounts to 3–4% for both buyer and seller.

5.4.3.2 Demand Uncertainty in the Spot Market

Having kept the relative degree of demand uncertainty constant in the previous section, the analysis now turns to studying *ceteris paribus* the impact of demand uncertainty in the spot market on the pricing and performance of long-term capacity contracts. Uncertainty of spot market demand is expressed by demand variance $\sigma^2_{\tilde{D}_S}$.

Optimal Pricing and Reservation Policies

Compared to the comparative static analyses in preceding sections, demand variance $\sigma^2_{\tilde{D}_S}$ seems to have little impact on the optimal pricing policy of the asset provider (see Fig. 5.25). When spot demand uncertainty increases, the capacity prices w^* and $r^* + x^*$ increase only slightly. Increasing uncertainty is unfavorable for the asset provider since the downside potential,

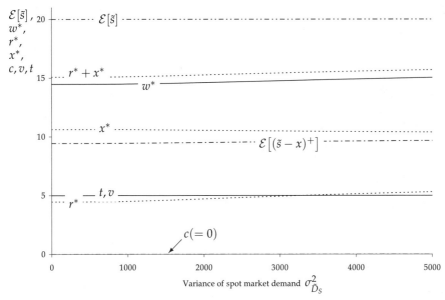

Fig. 5.25. Optimal pricing policy as a function of demand uncertainty in the spot market: As uncertainty in terms of demand variances increases, the asset provider shifts a greater portion of the total price to the certain price component r^*.

i.e., the risk of not finding enough buyers in the spot market to fill capacity, increases without limit, while the upside potential, i.e., the chance of benefiting from a high number of spot market buyers, is limited by the fixed amount of capacity. As in the analyses before, the total sum of reservation fee r^* and execution fee x^* of the optimal option contract exceed the capacity price w^* under the fixed-commitment contract because the option buyer can expect not to exercise all options. The evolution of r^* and x^* over $\sigma^2_{\tilde{D}_S}$ confirms the inference drawn in the discussion of the optimal pricing policy in Sec. 5.4.3.1: the lower the variance, the higher the optimal execution fee (though the effect here is less strong). With increasing uncertainty with regard to market demand, the asset provider shifts a greater portion of the total capacity to the certain part of the split tariff, namely the reservation fee.

Expected Profits and Welfare Analysis

While it has been discussed in Sec. 5.4.3.2 that the amount of reserved capacity *increases* when demand variance in the *contract* market increases, the optimal reservation policy depicted in Fig. 5.26 shows that the optimal amount of reserved capacity N^* *decreases* when demand variance in the *spot* market increases. The impact of demand variance is thus different, depending on whether contract or spot market demand is considered. The decrease of N^*

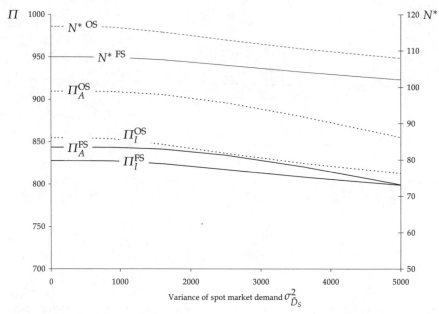

Fig. 5.26. Optimal reservation policy and expected profits as a function of spot demand uncertainty: The optimal numbers of reservation and expected profits decrease in demand variance.

in spot market demand variance is of indirect nature, namely a consequence of the higher price charged by the asset provider if spot price variance increases as discussed in the preceding subsection. The number of options sold $N^{* \, OS}$ exceeds the number of firm reservations $N^{* \, FS}$, again due to the higher service level induced by the option tariff. Since the reservation fee and thus the overage cost increase in $\sigma^2_{\tilde{D}_S}$, the difference in reserved volumes decreases in $\sigma^2_{\tilde{D}_S}$.

Independently of the type of contract used, Fig. 5.26 shows that the expected profits of both market parties decline in $\sigma^2_{\tilde{D}_S}$. Demand uncertainty in the spot market is thus *per se* unfavorable for capacity buyer and seller. Though option contracts yield higher expected profits for both buyer and seller for all levels of $\sigma^2_{\tilde{D}_S}$ (see Fig. 5.27), this advantage is shrinking as spot demand uncertainty increases. In contrast to the inferences drawn from Fig. 5.18 (the value of capacity options depends positively on demand uncertainty in the contract market), the value of capacity options correlates negatively with demand uncertainty in the spot market.

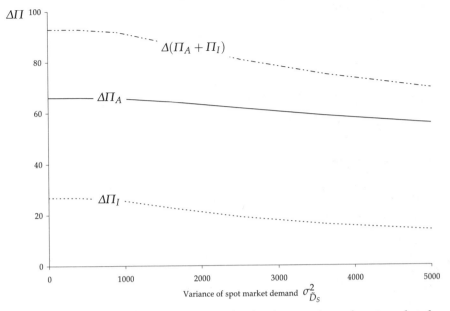

Fig. 5.27. Welfare comparison for varying levels of uncertainty of spot market demand: The difference in expected profits and thus the value of capacity options decrease in the demand variance.

5.4.3.3 Expected Spot Price

Drawing the analogy between capacity options as presented here and financial call options, e.g., on a company's stock, the spot price $\tilde{s}$ corresponds to the price of the underlying, i.e., the share price. The pricing of capacity options is affected by the expected value of the spot price $\mu_{\tilde{s}}$ and by the volatility of the underlying, which is represented by the spot price variance $\sigma_{\tilde{s}}^2$ (see Sec. 5.4.3.4).

As in the above discussion of spot market demand (see Sec. 5.4.3.1), the coefficient of variation $\vartheta_{\tilde{s}} = \sigma_{\tilde{s}}/\mu_{\tilde{s}}$ is kept constant during the following analysis at $\vartheta_{\tilde{s}} = 1/3$ in order not to change the relative level of uncertainty when varying the expected spot market price $\mu_{\tilde{s}}$.

Optimal Pricing and Reservation Policies

Fig. 5.28 shows the optimal pricing policy of the asset provider as a function of the expected value of the spot price $\mu_{\tilde{s}}$. The higher the expected spot price, the more expensive the asset provider sells capacity reservations. However, due to the price responsiveness of contract market demand, the prices w^* and $r^* + x^*$ only increase less than proportionally and converge for high values of $\mu_{\tilde{s}}$. If the expected spot price equals marginal cost (at $\mu_{\tilde{s}} = t = v = 5$), the asset provider earns zero margin if a fixed-commitment contract is

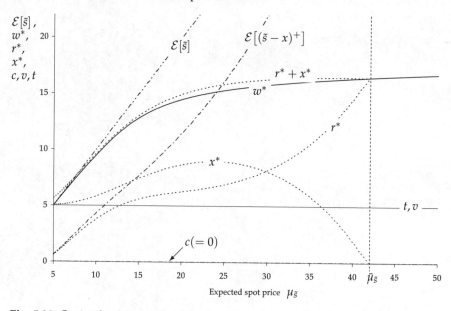

Fig. 5.28. Optimal pricing policy as a function of the expected spot price with constant coefficient of variation $\vartheta_{\tilde{s}} = 1/3$: When the expected spot price is greater than $\bar{\mu}_{\tilde{s}}$, the asset provider does not offer option contracts.

used. For the option contract in this case, the asset provider optimally sets $x^* = \mu_{\tilde{s}}$ and earns a positive margin from the reservation fee $r^* > 0$.

As the expected spot price increases, initially both the optimal reservation and execution fee increase; then, x^* decreases while r^* further increases until the execution fee has decreased to zero at $\mu_{\tilde{s}} = \bar{\mu}_{\tilde{s}}$. Since x^* cannot drop below zero, the asset provider only offers fixed-commitment contracts for $\mu_{\tilde{s}} \geq \bar{\mu}_{\tilde{s}}$.

Lemma 5.3. *The threshold value $\bar{\mu}_{\tilde{s}}$ is implicitly given by the solution to the following set of equations for $(r^*, \bar{\mu}_{\tilde{s}})$ with $x = 0$:*

$$N - \left[r^* - c - (\mu_{\tilde{s}} - t)(1 - \Phi(z_B)) \right] \left[\frac{\sigma_{\tilde{D}_C}}{\varphi(z_N)\mu_{\tilde{s}}} + b \right]$$
$$+ v \left[\frac{\sigma_{\tilde{D}_C}(1 - \Phi(z_N))}{\varphi(z_N)\mu_{\tilde{s}}} + b \right] = 0 \quad (5.13)$$

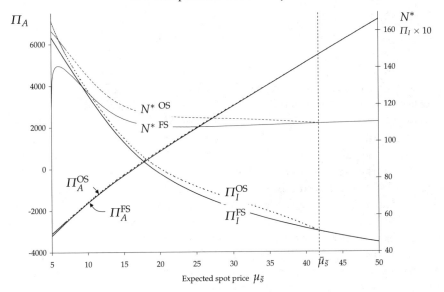

Fig. 5.29. Optimal reservation policy and expected profits as a function of the expected spot price with constant coefficient of variation $\vartheta_{\tilde{s}} = 1/3$: If the spot price is expected to be high, the intermediary reserves fewer capacity.

$$a - b(r^* + \lambda) - \sigma_{\tilde{D}_C} L(z_N)$$

$$- \left[r^* - c - (\mu \bar{s} - t)(1 - \Phi(z_B)) \right] \left[\frac{\sigma_{\tilde{D}_C}(1 - \Phi(z_N))}{\varphi(z_N)\mu \bar{s}} + b \right]$$

$$+ v \left[\frac{\sigma_{\tilde{D}_C}(1 - \Phi(z_N))^2}{\varphi(z_N)\mu \bar{s}} + b \right] = 0 \quad (5.14)$$

Proof. At $\bar{\mu}_{\tilde{s}}$, it is $r = r^*$ and $x = x^* = 0$ and thus $\partial \Pi_A / \partial r = \partial \Pi_A / \partial x = 0$ (first order condition, unconstrained optimization). Let $\mu_{\tilde{s}} = \bar{\mu}_{\tilde{s}}$. Taking $\partial \Pi_A / \partial r$ from (4.30) with (4.31)–(4.33) and setting $\partial \Pi_A / \partial r = 0$ and $x = 0$ gives (5.13). Taking $\partial \Pi_A / \partial x$ from (4.34) with (4.35)–(4.36) and setting $\partial \Pi_A / \partial x = 0$ and $x = 0$ gives (5.14).   □

The steady increase of r^* for $\mu_{\tilde{s}} < \bar{\mu}_{\tilde{s}}$ results from the fact that the intermediary's procurement alternative, i.e., the spot market, becomes more expensive when $\mu_{\tilde{s}}$ increases, allowing the asset provider to shift a higher portion of his total price to the upfront payable price component r.

Expected Profits and Welfare Analysis

The optimal reservation policy is displayed in Fig. 5.29. Generally, the optimal number of reservations declines in the expected spot price. Intuitively, one might have expected a different result: Since it is the intermediary's alternative procurement market that is becoming more expensive when the expected spot price increases, one might expect the intermediary to shift more of his capacity procurement to the contract market and thus reserve more capacity. However, this behavior of N^* can only be observed for fixed-commitment contracts ($N^{*\,FS}$) at values of $\mu_{\tilde{s}}$ slightly above the asset provider's variable cost. For all other values of $\mu_{\tilde{s}}$ and for the option contract without exception, this effect is offset by an opposite effect: Since the asset provider also raises the price in the contract market – which has a negative impact on the demand faced by the intermediary due to price responsiveness of demand – the number of reservations generally declines in $\mu_{\tilde{s}}$. As long as the asset provider offers option contracts, the number of options $N^{*\,OS}$ exceeds the number of firm reservations $N^{*\,FS}$ due to the higher optimal service level resulting from lower underage cost.

The expected profits (see Fig. 5.29) of intermediary and asset provider are adversely affected by an increase of the expected spot price. The asset provider's expected profits strongly increase in $\mu_{\tilde{s}}$ since, at constant cost, a higher spot price means a higher margin. The intermediary's profits decline in $\mu_{\tilde{s}}$ since the spot price represents a part of his procurement cost. However, the impact is much less pronounced than for the asset provider since the prices for contracted capacity (the other part of the intermediary's procurement cost) increase less than the spot price.

If the asset provider offers option contracts, which is true for all levels of $\mu_{\tilde{s}}$ between variable cost and $\bar{\mu}_{\tilde{s}}$, both buyer and seller benefit as compared to using a fixed-commitment contract because their expected profits increase (see Fig. 5.30). Option contracts are thus Pareto improving with respect to variations of the expected spot price. For the asset provider, the relative profit improvement is low for high values of $\mu_{\tilde{s}}$ since his absolute expected profit is high and the improvement only marginal. However, the relative impact is high for values between approximately $\mu_{\tilde{s}} = 15$ and $\mu_{\tilde{s}} = 20$ for which the asset provider operates at the edge of profitability (see also Fig. 5.29). The relative improvement of the intermediary's expected profit amounts to up to 7.5%.

5.4.3.4 Price Uncertainty in the Spot Market

Optimal Pricing and Reservation Policies

The uncertainty of the spot price is represented by the spot price variance $\sigma_{\tilde{s}}^2$. The price of capacity w^* under the fixed-commitment contract is independent of the spot price variance $\sigma_{\tilde{s}}^2$ (see Fig. 5.31). Since reservations under this type of contract are firm and payments sunk, the buyer will claim his

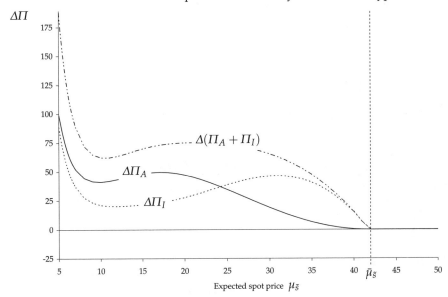

Fig. 5.30. Welfare comparison for different levels of expected spot price with constant coefficient of variation $\vartheta_{\tilde{s}} = 1/3$: Option contracts are Pareto improving with respect to variations of the expected spot price.

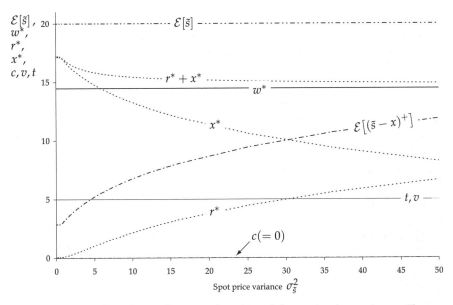

Fig. 5.31. Optimal pricing policy as a function of the spot price variance: The capacity price w^* under the fixed-commitment contract is constant, while the optimal reservation fee r^* and execution fee x^* change with spot price variance.

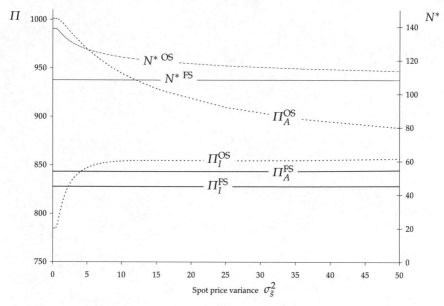

Fig. 5.32. Optimal reservation policy and expected profits as a function of the spot price variance: An increase of spot price variance comes at the cost of the asset provider.

capacity independent of the realization of the spot price; it is only the expected value of the spot price that determines the optimal capacity price. For the option contract, the optimal reservation and execution fee do depend on the uncertainty of the spot price because it is the spot price variance that determines the chance of the actual spot price being below or above the execution fee and thus the chance of option execution or non-execution. Consequently, if $\sigma_{\tilde{s}}^2$ increases, the probability of x being smaller than $\tilde{s}$ *ceteris paribus* increases. The asset provider's optimal execution fee x^* thus decreases in $\sigma_{\tilde{s}}^2$, which reduces the (from the asset provider's perspective) risk of non-execution. The asset provider recoups the margin loss associated with the decline of x^* by an increase of the optimal reservation fee r^* in $\sigma_{\tilde{s}}^2$. The total price $r^* + x^*$ declines in $\sigma_{\tilde{s}}^2$ and is generally higher than w^*.

The optimal reservation policy depicted in Fig. 5.32 shows that the number of reservations $N^{*\,\text{FS}}$ bought under the fixed-commitment contract is also, as a direct consequence of the flat capacity price w^*, independent of $\sigma_{\tilde{s}}^2$. The optimal number of options $N^{*\,\text{OS}}$ is greater than the number of firm reservations for all levels of $\sigma_{\tilde{s}}^2$. The more the reservation fee r^* increases, the more the intermediary's optimal service level and thus the number of options $N^{*\,\text{OS}}$ converges to the level under the fixed-commitment contract.

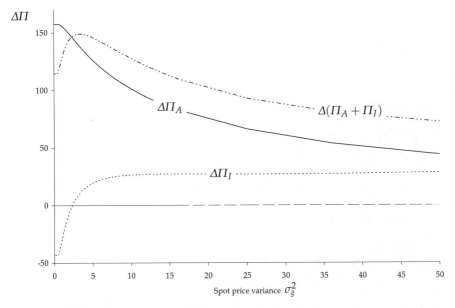

Fig. 5.33. Welfare comparison for different levels of spot price variance: An increase of spot price variance comes at the cost of the asset provider.

Expected Profits and Welfare Analysis

The expected profits (see Fig. 5.32) under the fixed-commitment contract are also independent of $\sigma_{\tilde{s}}^2$, as to be expected having observed the optimal pricing and reservation policies. As a consequence of the decline of total capacity price $r^* + x^*$ in $\sigma_{\tilde{s}}^2$, the intermediary's expected profit under the option contract increases in $\sigma_{\tilde{s}}^2$, the asset provider's expected profit decreases. An increase of the spot price variance comes at the cost of the asset provider.

The asset provider benefits from the option contract most when spot price uncertainty is low (see Fig. 5.33). The profit advantage of the option contract becomes smaller when the spot price uncertainty is higher, but is always positive. The intermediary's benefit increases in $\sigma_{\tilde{s}}^2$, and thus a higher spot price variance is especially favorable for the intermediary. However, for low degrees of spot price uncertainty, the intermediary is worse off when options are used instead of firm reservations. This might lead to conflicts, given the Stackelberg structure of the model: Since – especially for low degrees of spot price uncertainty – the asset provider is better off if option contracts are used, he being the Stackelberg leader will prefer to offer a capacity-option contract to the intermediary who, however, would prefer a fixed-commitment contract and probably demand compensation (e.g., by a side payment) when asked by the asset provider to switch from a fixed-commitment to a capacity-option contract.

The behavior of capacity options with regard to spot price uncertainty is in line with the notion of financial options: The higher the volatility of the underlying the higher (from the buyer's perspective) the value of the option.

5.4.4 Cost Structure

So far it has been assumed that the cost for capacity supply in the contract and spot market is identical (i.e., $c + v = t$). For the contract market, it has further been assumed that the entire cost for capacity supply does not arise until the reserved capacity is actually called on (i.e., reservation cost $c = 0$). These assumptions will now be relaxed. In the following, analyses are conducted with respect to

1. the level of variable cost under the assumption that variable cost is identical in contract and spot market;
2. the level of variable cost under the assumption that variable cost in the contract market differs from the cost in the spot market;
3. the level of reservation cost assuming that the variable cost in the contract market is incurred partly when capacity is reserved (reservation cost c) and partly at execution (variable cost v), with the total variable cost ($c + v$) in the contract market being identical to the variable cost in the spot market (t).

5.4.4.1 Identical Variable Cost in Contract and Spot Market

In this first analysis with regard to the asset provider's cost structure, the base-case assumption of identical variable cost in the contract and spot market is maintained. This section studies how the optimal reservation and pricing policies as well as the expected profits depend on the level of variable cost. To this end, the parameters v (variable cost in the contract market) and t (variable cost in the spot market) are changed simultaneously so that always $v = t$.

Optimal Pricing and Reservation Policies

The capacity prices w^* and $r^* + x^*$ generally increase as variable cost increases, however, the increase is less than proportional (see Fig. 5.34). In the case of the fixed-commitment contract, the optimal capacity price w^* does not exceed the expected spot price since this is the participation constraint of the intermediary. In case of the option contract, the sum of reservation and execution fee does exceed the expected spot price for high cost levels. Initially, both the optimal reservation fee r^* and the optimal execution fee x^* increase in variable cost. This implies a decline of the expected spot

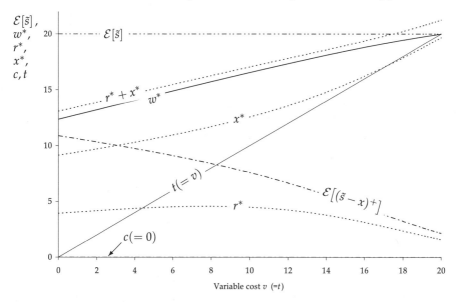

Fig. 5.34. Optimal pricing policy as a function of variable cost if variable cost is identical in contract and spot market: The total price $r^* + x^*$ for capacity under the option contract rises above the expected spot price if variable cost is high.

market premium $\mathcal{E}[(\tilde{s} - x)^+]$, which has to exceed the reservation fee r for the intermediary to participate in the option market. Therefore, the optimal reservation fee r^* then decreases as v further increases.

The execution fee x^* exceeds the variable cost v for most values of v; only for values of v close to the expected spot price, the execution fee is not cost-covering, i.e., remarkably, the asset provider optimally sets the execution fee even *below* variable execution cost. The expected loss from option execution thus must – for these values of v – be smaller than the benefit gained from stimulating demand and reservations through the optimal tariff.

Expected Profits and Welfare Analysis

The optimal reservation policy in response to the above outlined optimal pricing policy is shown in Fig. 5.35. Generally, the optimal number of capacity reservations decline in variable cost because the increase of variable cost is accompanied by a price increase. The two contract types differ with regard to the extent of the decline of the number of reservations. Under the fixed-commitment contract, the optimal number of reservations declines stronger and drops to zero as variable cost and thus the optimal capacity price w^* reaches the level of the expected spot price. In contrast, options are sold even at this cost level since the intermediary's participation constraint $r < \mathcal{E}[(\tilde{s} - x)^+]$ is still met.

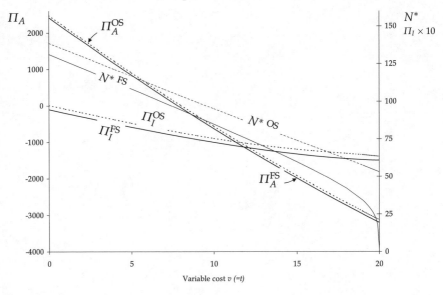

Fig. 5.35. Optimal reservation policy and expected profits as a function of variable cost if variable cost is identical in contract and spot market: As variable cost rise, the number of capacity reservations under the fixed-commitment contract drops to zero.

The expected profits (see Fig. 5.35) of both market players decline in variable cost. The intermediary's profits, however, are much less affected than the asset provider's (note that Π_A is because the intermediary, on the one hand, partly shifts his capacity procurement to the spot market while, on the other hand, the asset provider can only pass on a fraction of the cost increase to the intermediary via an increase of capacity prices. Consequently, the asset provider's profit decline strongly and monotonously in variable cost.

If an option contract is used, both parties expect higher profits for all levels of variable cost (for $0 \leq v \leq \mathcal{E}[\tilde{s}]$) as compared to the fixed-commitment contract (see Fig. 5.36). The absolute profit increase is higher for the asset provider than for the intermediary and is relatively constant for the levels of variable cost considered here. The value of the option contract is thus only marginally influenced by the height of variable cost for the case of identical cost in contract and spot market. However, one can observe in Fig. 5.36 that the incremental expected profits for both market participants exhibit an increase for high levels of variable cost. This is due to the fact that the number of reservations under the fixed-commitment contract converges to zero when v approaches $\mathcal{E}[\tilde{s}]$, while, in case of the capacity-option contract, the intermediary still reserves capacity at this cost level, resulting from the above described optimal pricing policy of the asset provider.

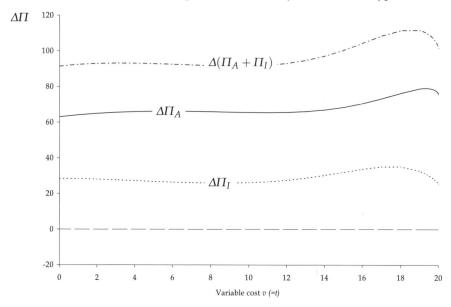

Fig. 5.36. Welfare comparison for different levels of variable cost if variable cost is identical in contract and spot market: The profit differences between option contract and fixed-commitment contract is only influenced marginally by the level of variable cost.

5.4.4.2 Different Variable Cost in Contract And Spot Market

The analysis now relaxes the assumption that variable cost are identical in contract and spot market. Variable cost in the contract market are denoted by v, variable cost in the spot market by t. In the following analysis, t is held constant at $t = 5$ while v takes on values from 0 to 5. The case of $v \leq t$ is particularly important since it is probably the one most often observed in reality. Sales in the contract market imply early information about market demand which the asset provider possibly can turn into a cost advantage that results from optimized route planning and allocation of transport assets.

Optimal Pricing and Reservation Policies

Fig. 5.37 depicts the optimal pricing policy of the asset provider as a function of the variable cost in the contract market v. The lower this variable cost of "production" v is, the lower the capacity prices w^* and $r^* + x^*$. However, the optimal prices go up only less than proportional when v increases, i.e., the asset provider can pass on the cost increase only partly due to the price responsiveness of contract market demand. In the case of the capacity-option contract, the majority of the price increase in v takes places in the form of

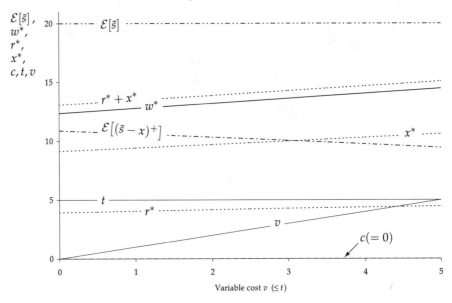

Fig. 5.37. Optimal pricing policy as a function of variable cost in the contract market v with constant variable cost in the spot market t: The asset provider passes on a change in variable cost only partly to the intermediary.

the optimal execution fee x^*. The optimal reservation fee r^* only increases slightly in v because the cost v is only incurred at option execution.

Expected Profits and Welfare Analysis

The optimal reservation policy of the intermediary is shown in Fig. 5.38. The optimal amounts of capacity reservations N^* generally decline in v, because the asset provider's optimal tariff increases in v. As in the above analyses, the optimal number of options $N^{*\ OS}$ is generally higher than the optimal number of firm reservations $N^{*\ FS}$.

The expected profits of both parties also decline in v (see also Fig. 5.38). The asset provider's expected profits decline more strongly than the intermediary's since the asset provider can pass on an increase of variable cost only partly to the intermediary (see above). For both the asset provider and the intermediary, the profits expected when using an option contract are higher than when using a fixed-commitment contract.

Comparing the expected profits directly (see Fig. 5.39) shows that the absolute profit advantage from using option contracts hardly changes when the variable cost v changes. The possibility to achieve a Pareto improvement is independent of the level of variable cost and thus of the cost difference between long-term (contract market) and short-term capacity allocation (spot market).

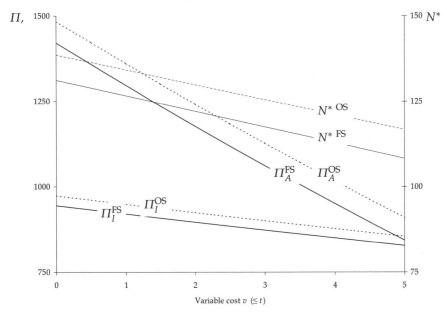

Fig. 5.38. Optimal reservation policy and expected profits as a function of variable cost in the contract market v with constant variable cost in the spot market t: Independent of the type of contract, expected profits and optimal number of reservations decline in v.

5.4.4.3 Reservation Cost in Contract Market

Finally, the assumption that the entire variable cost of capacity reserved in the contract market is incurred not before the actual use of the reserved capacity is relaxed. In the following analysis it is assumed that the variable cost in the contract market exists of two components of which one, the (variable) reservation cost c, is incurred at the moment the capacity is reserved and the other, the variable cost v, at the moment the capacity is used. This applies to both the option and the fixed-commitment contract.

Furthermore, it is assumed that the total variable cost in the contract market is equal to the variable cost in the spot market, i.e, $c + v = t$. In the following, an increase of c thus comes along with an decrease of v.

Optimal Pricing and Reservation Policies

The total prices for capacity w^* and $r^* + x^*$ are hardly affected by shifts in the structure of variable cost in the contract market (see Fig. 5.40). They increase only slightly the higher the cost component c, which is incurred at the moment of reservation. The total price under the option contract $r^* + x^*$ is generally higher than the price w^* under the fixed-commitment contract. In line with the observations in Sec. 5.4.4.2, the optimal execution fee x^*

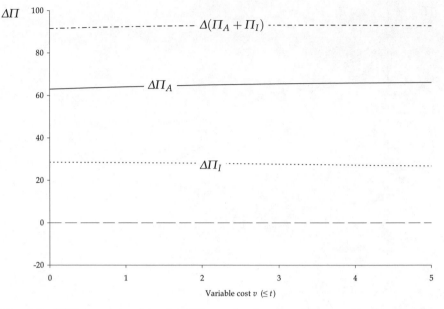

Fig. 5.39. Welfare comparison for different levels of variable cost in the contract market v with constant variable cost in the spot market t: The profit advantage of option contracts is hardly affected by a cost difference between spot and contract market.

decreases in c and thus increases – as above – in v. The optimal reservation fee r^* increases in c. The higher the cost component incurred upfront, the higher the price component charged upfront and analogously for the price/cost component incurred at execution. However, the price movements are always less than proportional to the cost changes. The asset provider thus passes on cost changes only partly to the intermediary.

Expected Profits and Welfare Analysis

The optimal reservation policy is depicted in Fig. 5.41. For both contract types, the optimal numbers of reservations N^* decline in c because the higher the reservation cost c becomes, the greater the sunk portion of the total variable cost. Though the optimal number of options $N^{*\ OS}$ is generally higher than the optimal number of reservations $N^{*\ FS}$ under the fixed-commitment contract, $N^{*\ OS}$ declines stronger in c than $N^{*\ FS}$ as the asset provider adjusts the structure of the split tariff and shifts a higher portion of the total price towards the upfront reservation fee r. The expected profits (see Fig. 5.41) of the two parties likewise decline in c and do so stronger in the case of the option contract.

Comparing the expected profits among the two contracts types shows that the option contract always results in a positive profit increment for

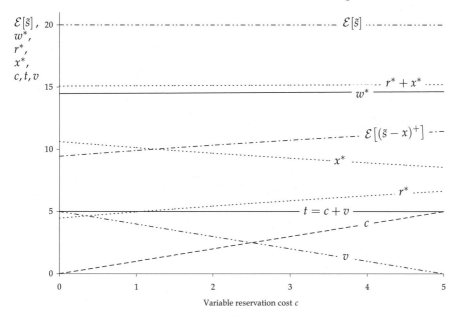

Fig. 5.40. Optimal pricing policy as a function of reservation cost c with constant total variable cost $c + v$: The price components of the split tariff under the option contract move in the same direction as the respective components of variable cost, but less than proportionally.

both the asset provider and the intermediary (see Fig. 5.42). The relative benefits amount to up to 7% in expected profits for the asset provider and up to 2.6% for the intermediary over the range of c displayed in Fig. 5.42. However, the profit surplus is decreasing in c. The higher the sunk part of total variable cost, the smaller the advantage of the option contract over the fixed-commitment contract.

5.5 Reduction of Double-Marginalization

So far, the performance improvement resulting from option contracts has been measured in absolute terms or relative to fixed-commitment contracts. As elaborated on in Chap. 3, double marginalization introduces inefficiencies in supply chains. The extent to which double marginalization occurs depends, among other things, on the type of supply contract used by the supply chain partners. The efficiency of a contract type can be assessed by measuring the extent of double marginalization. The maximum efficiency is reached if no double marginalization occurs. The supply chain is then said to be coordinated (channel coordination). This is the case when the contract parties act like one single decision maker (in the following called

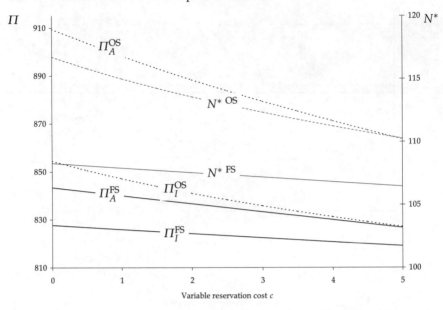

Fig. 5.41. Optimal reservation policy and expected profits as a function of reservation cost c with constant total variable cost $c + v$: In case of the option contract, the optimal number of reservations and expected profits decline stronger in c than for the fixed-commitment contract.

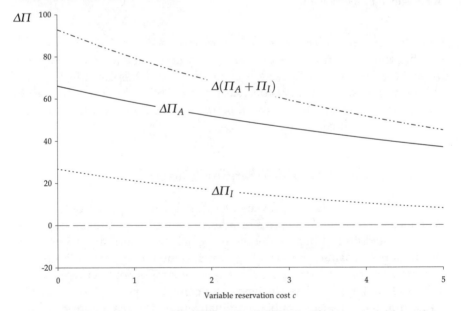

Fig. 5.42. Welfare comparison for different levels of reservation cost c with constant total variable cost $c + v$: When the variable cost per reservation c increases (holding $t = c + v$ constant), the profit advantage of the option contract becomes smaller.

integrated firm) who optimizes overall profit. In the following, at first, the optimal policy of the integrated firm is derived to establish the theoretical channel optimum. Then, the channel performance under the two contract types considered in the capacity-option pricing model is measured against this benchmark.

The integrated firm is the fiction of the asset provider and intermediary being one single entity, i.e., as if the asset provider served directly the intermediary's end-customers. Ultimately, this means that the integrated firm serves two kinds of customers. To the first group, the integrated firms sells at the fixed price p. The demand $\tilde{D}_C$ of this group is, as above, price-sensitive and its entire demand must be served by the integrated firm. Therefore, the fictitious integrated firm needs to "reserve" capacity for this customer group with itself. Again, the reserved capacity is denoted by N. If the integrated firm has reserved less capacity than this customer group demands ($N < \tilde{D}_C$), it purchases additional capacity at the uncertain price $\tilde{s}$ from other capacity providers on the spot market. This at first somewhat implausible assumption is necessary to reflect the setup of the original model, where it has been assumed that the intermediary purchases capacity from any arbitrary spot market seller and not necessarily from the asset provider. To the second group with demand $\tilde{D}_S$, the integrated firm can sell any non-reserved capacity, i.e., $K - N$ at the uncertain (spot) price $\tilde{s}$.

Except for the prices p, r, x, and w, the definition of and assumptions about all other variables are identical to the definitions and assumptions in Chap. 4. There, the prices r, x, and w denote the prices charged by the asset provider from the intermediary and are thus not applicable in the integrated channel. Instead, the integrated firm now directly sets p.

Let P_G denote the profit earned by the integrated firm:

$$P_G = p\tilde{D}_C - cN - v\min(\tilde{D}_C, N) - \tilde{s}(\tilde{D}_C - N)^+ + (\tilde{s} - t)\min(B, \tilde{D}_S) - fK \tag{5.15}$$

$$\text{with} \quad B = K - N$$

The objective of the integrated firm is to maximize expected profit $\mathcal{E}[P_G] = \Pi_G$ by optimally choosing the number of reservations N and the price p.

$$\max_{p,N} \Pi_G = \max_{p,N} \mathcal{E}\left[p\tilde{D}_C - cN - v\min(\tilde{D}_C, N) - \tilde{s}(\tilde{D}_C - N)^+ \right.$$
$$\left. + (\tilde{s} - t)\min(B, \tilde{D}_S) - fK \right] \tag{5.16}$$

The expected profit of the integrated firm is given by

$$\Pi_G = p\mathcal{E}\left[\tilde{D}_C\right] - cN - v\left[\int\limits_0^N \tilde{D}_C f_C(\tilde{D}_C)\mathrm{d}\tilde{D}_C + \int\limits_N^\infty N f_C(\tilde{D}_C)\mathrm{d}\tilde{D}_C\right]$$

$$- \mathcal{E}[\tilde{s}]\int\limits_N^\infty (\tilde{D}_C - N)f(\tilde{D}_C)\mathrm{d}\tilde{D}_C$$

$$+ (\mathcal{E}[\tilde{s}] - t)\left[\int\limits_0^{K-N} \tilde{D}_S f_S(\tilde{D}_S)\mathrm{d}\tilde{D}_S + \int\limits_{K-N}^\infty (K-N)f_S(\tilde{D}_S)\mathrm{d}\tilde{D}_S\right] - fK \quad (5.17)$$

and can be calculated analytically by

$$\Pi_G = p\mu_{\tilde{D}_C} - cN - v(\mu_{\tilde{D}_C} - \sigma_{\tilde{D}_C}L(z_N))$$
$$- \mu_{\tilde{s}}\sigma_{\tilde{D}_C}L(z_N) + (\mu_{\tilde{s}} - t)(\mu_{\tilde{D}_S} - \sigma_{\tilde{D}_S}L(z_B)) - fK. \quad (5.18)$$

Theorem 5.1. *The integrated firm's optimal choice (N^*, p^*) satisfies the set of equations:*

$$c - [1 - \Phi(z_{N^*})](\mu_{\tilde{s}} - v) + [1 - \Phi(z_B)](\mu_s - t) = 0 \quad (5.19)$$

$$\mu_{\tilde{D}_C} + (v - p^*)b - \left[(\mu_{\tilde{s}} - t)[1 - \Phi(z_B)] + c\right]\frac{\partial N}{\partial p}$$
$$+ (\mu_{\tilde{s}} - v)[1 - \Phi(z_{N^*})]\left(\frac{\partial N}{\partial p} + b\right) = 0 \quad (5.20)$$

with

$$\frac{\partial N}{\partial p} = -\frac{(\mu_{\tilde{s}} - v)\frac{\varphi(z_N)}{\sigma_{\tilde{D}_C}}b}{(\mu_{\tilde{s}} - v)\frac{\varphi(z_N)}{\sigma_{\tilde{D}_C}} + (\mu_{\tilde{s}} - t)\frac{\varphi(z_B)}{\sigma_{\tilde{D}_S}}} \quad (5.21)$$

Proof. Differentiating Π_G from (5.17) with respect to N by applying Leibniz' rule, rearranging terms, setting $\partial\Pi_G/\partial N = 0$, and applying the relation in (A.8) (first-order condition) gives (5.19). With

$$\frac{\partial\Phi(z_N)}{\partial p} = \frac{\varphi(z_N)}{\sigma_{\tilde{D}_C}}\left(\frac{\partial N}{\partial p} + b\right) \quad \text{and} \quad (5.22)$$

$$\frac{\partial\Phi(z_B)}{\partial p} = -\frac{\varphi(z_B)}{\sigma_{\tilde{D}_S}}\frac{\partial N}{\partial p}, \quad (5.23)$$

differentiating (5.19) with respect to p and solving for $\partial N/\partial p$ gives (5.21) (implicit differentiation). With

Table 5.2. Extent of double marginalization in the base case

	N^*	Π_I	Π_A	$\Pi_I + \Pi_A$	Π_G	Percentage[a]
Fixed-commitment contract	108.29	827.72	843.38	1671.10	n/a	87.8%
Capacity-option contract	116.73	854.50	909.49	1763.99	n/a	92.7%
Integrated firm	163.93	n/a	n/a	n/a	1903.44	100.0%

[a] Joint expected profit $\Pi_I + \Pi_A$ relative to integrated firm Π_G.

$$\frac{\partial L(z_N)}{\partial p} = -\frac{1 - \Phi(z_N)}{\sigma_{\tilde{D}_C}} \left(\frac{\partial N}{\partial p} + b \right) \quad \text{and} \tag{5.24}$$

$$\frac{\partial L(z_B)}{\partial p} = \frac{1 - \Phi(z_B)}{\sigma_{\tilde{D}_S}} \frac{\partial N}{\partial p}, \tag{5.25}$$

differentiating Π_G from (5.18) with respect to p, simplifying, rearranging terms, and setting $\partial \Pi_G / \partial p = 0$ (first-order condition) gives (5.20). $\qquad \square$

Given the form of (5.19) and (5.20), N^* and p^* can be determined by applying numerical solution methods (see Sec. 5.1.3).

Applying the optimal policy derived in Theorem 5.1, the expected profit of the integrated firm can be calculated and compared to the sum of the expected profits (*joint expected profit* in the following) of asset provider and intermediary when using fixed-commitment or capacity-option contracts. Table 5.2 shows for the base case that the channel forfeits 12.2% of the maximum theoretical expected profit when a fixed-commitment contract is used, and only 7.3% for the case of a capacity-option contract. Though neither option contracts nor fixed-commitment contracts can fully coordinate the supply chain, capacity-option contracts are able to reduce double marginalization by 5 percentage points and thus increase channel efficiency.

5.6 Summary of Results

Summarizing the preceding results of the comparative static analysis, it has been shown that option contracts dominate fixed-commitment contracts in most cases. Except for the following three cases, option contracts yield higher expected profits for the asset provider than fixed-commitment contracts:

- The contract market is very small (relative to total capacity), i.e., $a < \bar{a}$ (see Lemma 5.1),
- or price-sensitivity in the contract market is very high, i.e., $b > \bar{b}$ (see Lemma 5.2),

- or the expected spot price is very high, i.e., $\mu_{\tilde{s}} > \bar{\mu}_{\tilde{s}}$ (see Lemma 5.3).

In these cases, the asset provider does not offer option contracts but reverts to fixed-commitment contracts. In the cases, where option contracts are chosen, they regularly make *both* asset provider and intermediary better off, i.e., a Pareto improvement is achieved. However, two exceptions have been found in the case of which the intermediary expects lower profits than he would expect if a fixed-commitment contract were offered:

- The spot price variance is very low, i.e., $\sigma_{\tilde{s}}^2 \to 0$ (see Sec. 5.4.3.4)
- or capacity is scarce, i.e., K around $\mu_{\tilde{D}_C} + \mu_{\tilde{D}_S}$ (see Sec. 5.4.1).

The welfare loss for the intermediary in the first case is considerably large (up to 5.2% of expected profits), in the latter case it is rather small (less than 0.7% of expected profits). In both cases, the joint profit improvement is always positive. Hence even in these cases, a Pareto improvement could possibly be achieved, e.g., by the asset provider compensating the intermediary through a side-payment.

Furthermore, it is doubtful if either of these conditions apply in reality. With regard to the spot price variance, this is rather unlikely. As indicated in Sec. 2.1.2, spot prices are volatile and thus do exhibit a non-negligible variance. In fact, the quest for potentially lower spot prices has been found to be an important reason for the reluctance of forwarders to sign capacity agreements (see Sec. 2.2.4).

Capacity being of the order of magnitude of the sum of market demands may be more probable to apply in reality, since this is the equilibrium that – on average – balances market supply and demand. However, for the reasons outlined in Sec. 2.1.2, balancing demand and supply in the air cargo industry is aggravated by the fact that capacity supply is, for the case of belly capacity on passenger aircraft, triggered by the demand for passenger transport and that capacity cannot easily be adjusted at short notice and in large increments only. Furthermore, it has been discussed that, partly resulting from these issues, overcapacity is a frequently observed phenomenon in the industry, such that on many routes capacity and supply are hardly balanced. A negative – albeit minor – impact on the intermediary's expected profit may thus occur, but probably on those routes only, where supply and total demand is despite the obstacles mentioned above approximately balanced (for empirical findings on this issue, see Chap. 7).

Table 5.3 summarizes the findings about the determinants of the value of capacity option contracts. As before, $\Delta\Pi_A$ denotes the profit improvement expected for the asset provider ($\Delta\Pi_I$ for the intermediary) resulting from using capacity-option contracts instead of fixed-commitment contracts. The influence of the different sources of uncertainty on the value of capacity option contracts is especially noteworthy.

Table 5.3. Determinants of option contract value

An increase in ...	... lets	
	$\Delta\Pi_I$	$\Delta\Pi_A$
Capacity K	decrease/increase	increase
Contract market size a	increase/decrease	increase/decrease
Contract market demand variance $\sigma^2_{\bar{D}_C}$	increase	increase
Price sensitivity of contract market demand b	increase/decrease	increase/decrease
Expected spot market demand $\mu_{\bar{D}_S}$	decrease	decrease
Spot market demand variance $\sigma^2_{\bar{D}_S}$	decrease	decrease
Expected spot price $\mu_{\bar{S}}$	decrease	decrease
Spot price variance $\sigma^2_{\bar{S}}$	increase	decrease

The general notion about options is that uncertainty increases the value of an option (dating back to Merton 1973, who points out that the value of a financial option increases in the volatility of the underlying). The sources of uncertainty in the model are expressed by the variances of contract market demand, spot market demand, and spot price. The reaction of value of the capacity-option contract with regard to an increase in spot price variance corresponds to a financial option: for the option buyer (intermediary), the value of the option contract increases in spot price variance and decreases for the option seller (asset provider). An increase of spot market demand variance, however, is detrimental to both the capacity-option contract value for buyer and seller. In stark contrast, an increase in contract market demand increases the capacity-option contract value for both parties. The different sources of uncertainty thus differ substantially with regard to their respective effect on the value of the capacity-option contract. Similar findings about the necessity to distinguish between different sources of uncertainty and their respective effect on option value have been made by Huchzermeier and Loch (2001) with regard to real options in R&D projects.

Though the option contract value increases in some of the uncertainties discussed above, options are not a panacea against these uncertainties. The expected profits of the market participants regularly decline (or, at best, are flat) in the variances of the three uncertain variables no matter if option contracts or fixed-commitment contracts are being used. But the decline of expected profits is *lower* in the case of option contracts. The only exception constitutes the intermediary's expected profit which *increases* in spot price variance (see Sec. 5.4.3.4).

The main mechanism that makes a capacity-option contract yield higher expected profits than a fixed-commitment contract is the fact that it triggers a higher number of reservations. In general, the intermediary purchases options on more capacity units than he would otherwise – in the case of a fixed-commitment contract – reserve. This behavior essentially results from the cost trade-off in the Newsboy model that determines the intermediary's optimal service level and thus the optimal number of reservations. In the case of the option contract, the overage cost, i.e., the monetary amount that is lost for each unit of capacity the intermediary reserves ex ante in excess of his ex-post actual need, is constituted by the reservation fee r, in the case of the fixed-commitment contract by the capacity price w. Since r^* is generally not greater than w^*, the optimal service level of the intermediary increases and he reserves a greater amount of capacity (see Fig. 5.10), which is closer to the optimal number of reservations of the fictitious integrated firm introduced in Sec. 5.5 and thus reduces the extent of double marginalization.

The performance increase by using option contracts is *not* primarily a result of an increase of capacity utilization. On the contrary, capacity utilization may even decrease if option contracts are used (see Fig. 5.12). In general, two opposing effects exist with regard to the effect of option contracts on capacity utilization. On the one hand, capacity utilization increases since the asset provider sells more reservations and thus, in expectation, attracts more business from the intermediary. On the other hand, options may not be executed, leading to a decrease of capacity utilization. Which of these effects prevails depends on the situation at hand.

6

Model Extensions: Distribution of Profits, Correlations, and Overbooking

In Sec. 4.2, some assumptions have been made about the asset provider, the intermediary, and the markets they are acting in. These include risk neutrality of the market participants, independence of the spot market, and non-overbooking of reserved capacity. This chapter further analyzes the impact of these assumptions on the results of the model and partly relaxes them. Related to the assumption of market participants being risk neutral, Sec. 6.1 considers the distribution of expected profits; Sec. 6.2 deals with correlations between contract and spot market as well as between spot market price and spot market demand. Finally, Sec. 6.3 relaxes the non-overbooking assumptions and presents an overbooking model for option-based capacity reservations.

6.1 Distribution of Profits

In Sec. 4.2.1, it has been assumed that asset provider and intermediary are risk neutral and thus maximize *expected* profits without regard to the variability of profits. A main result in Chap. 5 has been that asset provider and intermediary prefer under most circumstances a capacity-option contract over a fixed-commitment contract because it yields higher expected profits for both parties. Relaxing the assumption about the attitude towards risk of the market participants can change this result, especially if the increase in expected profits comes at the expense of an increase in profit variability. Whether the players then still prefer the option contract, is determined by the shape of their respective utility function.

One might conjecture that the variability of the asset provider's profit increases when switching from a fixed-commitment to a capacity-option contract because the asset provider then faces in addition to price and demand uncertainty the risk that (some) options are not executed. The variability of

Table 6.1. Simulation statistics: Profit forecasts in the base case when the optimal policies are being followed.

| | Intermediary's profit $\tilde{P}_I$ | | Asset provider's profit $\tilde{P}_A$ | |
	Fixed-commitment contract	Capacity-option contract	Fixed-commitment contract	Capacity-option contract
Trials	10,000	10,000	10,000	10,000
Mean (analytical model)	830.4 (827.7)	853.9 (854.5)	841.9 (843.4)	912.7 (909.5)
Median	866.3	891.9	808.5	897.4
Standard deviation	219.1	194.6	1046.3	1089.2
Variance	48,013.3	37,860.4	1,094,734.2	1,186,439.6
Skewness	−1.16	−0.09	0.17	−0.13
Kurtosis	6.68	5.70	3.12	3.59
Coeff. of variation	0.26	0.23	1.24	1.19
Range minimum	−523.0	−23.8	−3234.0	−3052.5
Range maximum	2,089.6	2,269.1	5,371.1	5,317.4
Range width	2,612.6	2,292.9	8,605.2	8,369.9
Mean std. error	2.19	1.95	10.46	10.89

profits is analyzed in the following by the means of a Monte-Carlo simulation.

The simulation results presented here and in any following section have been derived using the spreadsheet simulation software *Crystal Ball 2000*, Version 5.2.2 by *Decisioneering Inc.*, Denver (Colo.) in combination with the spreadsheet software *Microsoft Excel*, Version 10 by *Microsoft Corp.*, Redmond (Wash.).

For the base case as specified in Table 5.1, Table 6.1 compares the distribution of profits earned by the asset provider and intermediary under both the fixed-commitment and capacity-option scenario. With 10,000 trials, the means of the sample profits approximately equal the expected profits of the analytical model presented in Chap. 4. As seen before (Sec. 5.4), both parties' expected profits are higher in the option-contract scenario. For the intermediary (see Fig. 6.1), the profit variance in the capacity-option scenario is lower than in the fixed-commitment scenario. Thus the intermediary does not only expect to earn more, but the higher profit exhibits also lower variability.

The profit variance for the asset provider (see Fig. 6.2), though, is higher in the capacity-option scenario than in the fixed-commitment scenario. However, the coefficient of variation (1.24 under the fixed-commitment con-

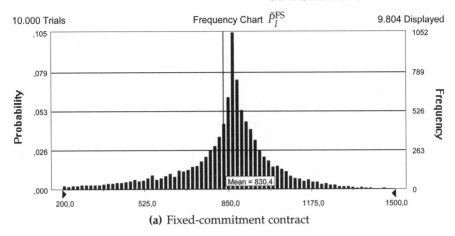

(a) Fixed-commitment contract

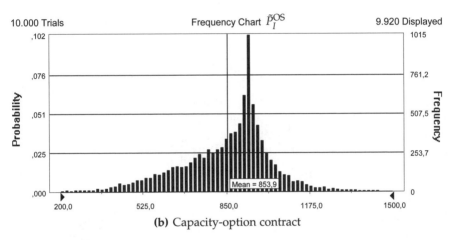

(b) Capacity-option contract

Fig. 6.1. Distribution of intermediary's profit $\tilde{P}_I$ (simulation results) in the base case when the optimal policies are being followed: The peak of the distribution shifts to the right, if an option contract is used.

tract, 1.19 for capacity options) shows that the riskiness of profit decreases for the asset provider, too. Besides this, Fig. 6.2(b) shows that the distribution of the asset provider's profit becomes bimodal. The second, lower peak at the left-hand tail of the profit distribution in the option-contract scenario results from non-execution of options due to spot price realizations below the execution fee.

In the base case, option contracts are thus for none of the market participants more risky (with regard to profit variability) than fixed-commitment contracts. However, this observation can only be an indication that the model results hold also in the case of attitudes towards risk different

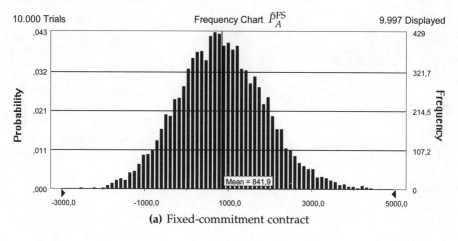

(a) Fixed-commitment contract

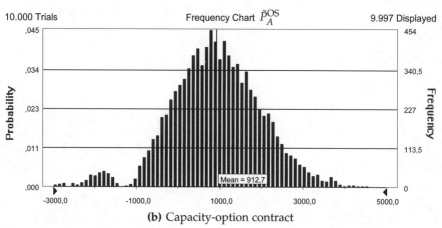

(b) Capacity-option contract

Fig. 6.2. Distribution of asset provider's profit $\tilde{P}_A$ (simulation results) in the base case when the optimal policies are being followed: The second, smaller peak at the left of the option-contract distribution results from options not being executed.

from risk neutrality, e.g., risk aversion. In order to draw general conclusions would require to carry out the analysis conducted in Chap. 4 and Chap. 5 on the basis of utility instead of profit maximization, which is beyond the scope of the capacity-pricing model presented here.

6.2 Interdependencies Between Stochastic Variables

The stochastic variables in the capacity-option pricing model – contract market demand $\tilde{D}_C$, spot market demand $\tilde{D}_S$, and spot price $\tilde{s}$ – have been assumed to be mutually independent (see Sec. 4.2.2 and 4.2.3). In the following, it is examined if these assumptions of independence are critical to the optimal choice of contract type, i.e., if the optimal choice of contract changes when these assumptions are relaxed.

6.2.1 Correlation Between Spot Market Demand and Spot Price

The optimal polices derived in Chap. 4 and analyzed in Chap. 5 have been determined under the assumption that spot market demand and spot price are independent. One can argue, though, that spot market demand and spot price are interdependent. In microeconomics, usually a negative dependence of price and demand is assumed: the higher the price, the fewer buyers are willing to pay that price and, thus, the lower is demand. This rationale characterizes, e.g., any negatively sloped demand function and will most certainly hold true for a market in total and in the long run (cf. Hyman 1993, p.170).[1] In the capacity-option pricing model, the portion of spot market demand $\tilde{D}_S$ faced by the asset provider has been assumed independent of the spot price $\tilde{s}$, with $\tilde{D}_S$ and $\tilde{s}$ being positive stochastic variables. Expected profits in the case of correlated spot market demand and spot price reported on in the following could therefore not be determined analytically from the model, but were estimated via simulation experiments.

Fig. 6.3 displays the results of simulation experiments with different coefficient of correlation between $\tilde{D}_S$ and $\tilde{s}$ (with all other variables as specified in the base case, see Table 5.1). It shows the impact on profits if market participants make their decisions assuming independence of $\tilde{D}_S$ and $\tilde{s}$ when indeed $\tilde{D}_S$ and $\tilde{s}$ are interrelated. The coefficient of correlation (cf. Rinne 1997, p. 369) is defined as

$$\rho_{\tilde{D}_S,\tilde{s}} \equiv \frac{\mathcal{E}\left[(\tilde{D}_S - \mu_{\tilde{D}_S})(\tilde{s} - \mu_{\tilde{s}})\right]}{\sigma_{\tilde{D}_S}\sigma_{\tilde{s}}}. \tag{6.1}$$

Each point in Fig. 6.3 represents the mean of profits achieved in 10,000 simulation trials with asset provider and intermediary following the optimal policies derived in Chap. 4, i.e., under the assumption of independence. Prices r, x, and w, respectively, as well as number of reservations N are thus constant within each the fixed-commitment and capacity-option scenario.

---

[1] In the short run, however, also positive dependance between spot price and demand may be observed. An individual seller may experience situations where he faces high demand and can charge a high spot price. For example, if demand is high and supply (capacity) fixed, the price in the spot market goes up. Vice versa, if demand is weak, the sellers may dump capacity on the market at a low price. Therefore, also positive correlations between spot market demand and spot price are reported on in the following.

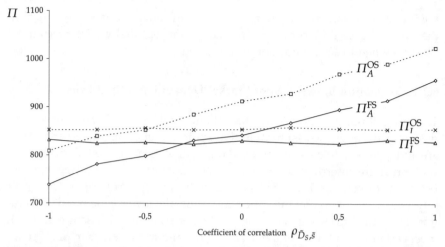

Fig. 6.3. Expected profits (simulation results) if spot market demand and spot price are correlated: the greater the coefficient of correlation, the higher the expected profits of the asset provider.

Fig. 6.3 shows that the expected profits of the intermediary Π_I do not depend on the correlation of $\tilde{D}_S$ and $\tilde{s}$ because Π_I is not a function of $\tilde{D}_S$ (see (4.4)). The expected profits of the asset provider increase in $\rho_{\tilde{D}_S,\tilde{s}}$ at a similar rate for both types of contract. This is because correlation of $\tilde{D}_S$ and $\tilde{s}$ determines the chances that an – from the perspective of the asset provider – unfavorable outcome of spot market demand, i.e., $\tilde{D}_S$ low, is offset by a favorable outcome of spot price, i.e., $\tilde{s}$ high. The larger $\rho_{\tilde{D}_S,\tilde{s}}$ is, the higher the chance that high values of $\tilde{D}_S$ occur together with high values of $\tilde{s}$, which improves the asset provider's profit.

Since Fig. 6.3 and similar simulation experiments for constellations different from the base base indicate that a correlation of spot market demand and spot price influence asset provider's expected profit, the possibility that the optimal pricing policy of the asset provider would also depend on $\rho_{\tilde{D}_S,\tilde{s}}$ cannot be ruled out. However, the analysis of this effect is beyond the scope of the capacity-option pricing model.

6.2.2 Demand Correlation Between Contract and Spot Market

Following the setup of the vast majority of revenue management models (cf. Weatherford and Bodily 1992; Tscheulin and Lindenmeier 2003), demands in contract and spot market have been assumed to be independently distributed.[2] One may argue, though, that demands are positively corre-

---

[2] See Mendelson and Tunca (2004) for an attempt to model a so-called closed market that considers such dependencies.

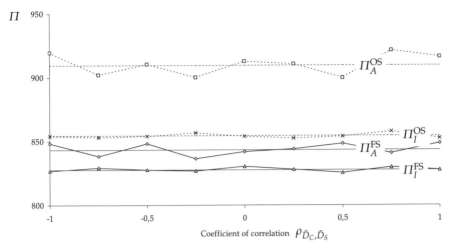

Fig. 6.4. Expected profits (simulation results) if demands in contract and spot market are correlated: The expected profits and thus the choice of contract type are not affected by the level of demand correlation.

lated because there might exist common factors that influence both contract and spot market demand, e.g., the state of the economy. Furthermore, if the structure on the demand side of the spot market is not perfectly competitive (as assumed here), but, e.g., rather represents an oligopsony (i.e., a limited number of capacity buyers), then demand captured by the asset provider in the contract market is likely to negatively affect demand in the subsequent spot market. For these reasons, it is examined in the following what impact demand correlation has on the expected profits of intermediary and asset provider and consequently on the optimal choice of contract type. Though of minor economic meaning, negative correlations are – for the completeness of the analysis – also included.

Fig. 6.4 shows the expected profits of asset provider and intermediary for different levels of correlation between demands in contract and spot market, measured by the coefficient of correlation $\rho_{\tilde{D}_C,\tilde{D}_S}$ which is defined by analogy with (6.1). As in Sec. 6.2.1, the points in Fig. 6.4 represent the results of simulation experiments with 10,000 trials each; the thin straight lines indicate the analytical value of expected profits in the base case with $\rho_{\tilde{D}_C,\tilde{D}_S} = 0.$[3]

---

[3] Note that the simulation results displayed in Fig. 6.4 fluctuate more widely around the analytically determined mean (thin straight lines) for the asset provider than for the intermediary. This is due to the fact that the simulation's mean standard error is generally larger (for the same number of trails) for the asset provider than for the intermediary (see Table 6.1).

The expected profits of the intermediary are independent of $\rho_{\tilde{D}_C, \tilde{D}_S}$ because Π_I is not a function of $\tilde{D}_S$. Likewise, the expected profits of the asset provider, which are functions of both $\tilde{D}_S$ and $\tilde{D}_C$, are not affected by correlation of demands. For the expected value as an average over all states it is irrelevant if high values of one demand variable are more likely to appear if the other demand variable takes on high (in the case of positive correlation) or low (negative correlation) values since both variables are realized at the same time.

Hence, no indication can be found that an interdependence of demands in contract and spot market affect the profits expected by asset provider or intermediary and thus the optimal choice of the contract type.

6.3 Overbooking

Intentionally setting booking levels higher than capacity, i.e., selling more capacity than one actually has, is called *overbooking* (cf. Smith et al. 1992). The booking level B as the maximum number of spot-market bookings the asset provider is willing to accept has so far been restricted to $B = K - N$: At most the non-reserved capacity has been sold in the spot market. However, if no-shows appear it can be valuable to overbook capacity, i.e., to set $B > K - N$. In the following, it is first defined what kind of no-shows are of particular relevance in the context of the market model considered here. Then, conditions under which the application of overbooking is useful are established and finally the optimal overbooking policy of the asset provider derived.

6.3.1 No-Shows Within the Capacity-Option Pricing Model

A no-show is a customer who has booked capacity but does not show up to make use of the capacity. Basically, no-shows can result from customers who have purchased capacity in the contract market or from customers who have booked in the spot market. Only the former will be considered here. Since the model focuses on contracting practices for capacity reservations, i.e., on the contract market, modeling of the spot market has been limited to what is necessary to provide asset provider and intermediary an alternative market to sell and buy capacity. Therefore, the model refrains especially from modeling no-show behavior of customers in the spot market (for a corresponding formulation with a stochastic show-up rate, see Kasilingam 1997) as this would yield no further insight about the optimal choice of contract type.

Aside from the not-modeled no-shows in the spot market, a different source of "no-shows" occurs when option contracts are used, namely the non-execution of options, which occurs for all options if the spot price $\tilde{s}$ is smaller than the execution fee x^* or for some options if demand of the intermediary $\tilde{D}_C$ is smaller than the number of options N^* (see Lemma 4.1). In these cases, the asset provider does not collect the execution fee x^* and thus loses revenue potential. Because it has been assumed in the model that fixed-commitment contracts can be enforced, no-shows associated with a revenue loss do not occur for reservations under a fixed-commitment contract because execution of reservation is free of charge for the intermediary ($x = 0$). Even if the actual demand of the intermediary $\tilde{D}_C$ is lower than the reserved quantity N^*, the asset provider still receives $w^* N^*$.

If no-shows occur and the asset provider has turned down spot-market demand because the number of spot market booking requests has exceeded the available capacity, he incurs – on the one hand – an opportunity cost. It will be shown below that the more capacity is overbooked, the lower this opportunity cost becomes. On the other hand, overbooking implies the risk that eventually more customers holding a reservation or booking show up than capacity is available. The asset provider then incurs an offload cost that results from compensations, contractual penalties, and loss of goodwill. This offload cost increases in the overbooking level (see also Fig. 3.4).

In the following it is analyzed to what extend a no-show problem exists if option contracts are used and how overbooking can be used to financially benefit from non-executed options.

6.3.2 Conditions for Useful Application

Overbooking can be useful in the model if the following conditions simultaneously apply:

1. The number of no-shows, i.e., non-executed options, is non-negligible. The expected number of non-executed options is $\mathcal{E}[N - \tilde{E}]$.
2. Spot demand $\tilde{D}_S$ exceeds the booking level B set by the asset provider, i.e., the number of denied spot-market bookings is positive. The expected number of denied spot market bookings is $\mathcal{E}[(\tilde{D}_S - B)^+]$

Fig. 6.5 displays the expected number of non-executed options and denied spot market bookings as a function of the expected spot market demand $\mu_{\tilde{D}_S}$ with all other variables at their base-case values (see Table 5.1). The expected number of non-executed options in the base case with $\mu_{\tilde{D}_S} = 200$ is $\mathcal{E}[N - \tilde{E}] = 11.96$. From this perspective, overbooking would be desirable to compensate for these expected no-shows. However, Fig. 6.5 also shows that for $\mu_{\tilde{D}_S} = 200$ the number of denied spot-market bookings $\mathcal{E}[(\tilde{D}_S - B)^+]$ is very close to zero. Thus, even if the capacity offered in the spot market were overbooked, it is very unlikely that the asset provider would find buyers for any additional capacity.

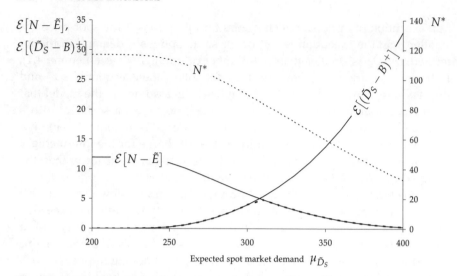

Fig. 6.5. Expected number of non-executed options $\mathcal{E}[N - \tilde{E}]$ and denied spot market bookings $\mathcal{E}[(\tilde{D}_S - B)^+]$ in the base case as functions of the expected spot market demand $\mu_{\tilde{D}_S}$.

In Fig. 6.5 overbooking is most promising for expected spot-market demand values around $\mu_{\tilde{D}_S} = 308$ because the thick dashed line indicating the minimum of non-executed options and denied spot-market bookings reaches its maximum value. For higher values of $\mu_{\tilde{D}_S}$, overbooking becomes less meaningful as the expected number of non-executed options converges to zero. This is because the optimal number of reservations N^* generally declines in $\mu_{\tilde{D}_S}$ (see Sec. 5.4.3.1).

Furthermore, the significance of overbooking is influenced by spot market demand uncertainty and spot price uncertainty: High spot market demand uncertainty *ceteris paribus* means a higher expected number of denied spot market bookings, while a high spot price uncertainty increases the chance of $\tilde{s} < x$ and thus the risk of non-execution of options.

6.3.3 Optimal Overbooking Policy

The overbooking model presented in the following follows at first the procedure of common revenue management systems that neglect the interdependence of the comprised optimization problems and solve them sequentially in order to reduce system complexity (cf. Stuhlmann 2000, p. 243, and Tscheulin and Lindenmeier 2003, p. 631). In line with this approach, the following formulation of the overbooking problem takes the solution to the reservation and pricing problems derived in Chap. 4 as given. In Sec. 6.3.5, it is shown how the optimal tariff would change if it were derived taking potential overbooking into account.

According to the sequence of events introduced in Chap. 4 and summarized in Fig. 4.2, the asset provider decides on the booking level B, i.e., the maximum number of spot market bookings he is willing to accept[4], during the first part (booking phase) of the spot market phase. He does so under uncertainty with respect to the number of executed options $\tilde{E}$ which is not resolved before the second part (execution phase) of the spot market phase.

For all capacity units that are called on either on the basis of spot market bookings $\tilde{M}$ or in the form of option executions $\tilde{E}$ in excess of total capacity K, thus for $\tilde{Q} \equiv (\tilde{M} + \tilde{E} - K)^+$ (in the following termed offload quantity), the asset provider incurs the offload cost g. Since $\tilde{M} = \min(\tilde{D}_S, B)$ as defined in Sec. 4.4, one gets $\tilde{Q} = \tilde{Q}(B)$. For the clarity of the exhibition, it is assumed that the offload cost g is constant[5].

The asset provider's objective is to maximize expected profits by optimally choosing B. The objective function from (4.19) is extended by the offload cost:

$$\max_B \Pi_A(B) - g\mathcal{E}\left[\tilde{Q}(B)\right]. \tag{6.2}$$

With the spot market margin $(\tilde{s} - t)$ representing the opportunity cost for capacity units that could have been sold in the spot market if they had not been blocked by reservations and Γ denoting total expected overbooking cost, the profit maximization problem in (6.2) can be formulated as the following corresponding cost minimization problem:

$$\min_B \Gamma(B) = \min_B \mathcal{E}\left[(\tilde{s} - t)(\tilde{D}_S - B)^+ + g\tilde{Q}(B)\right] \tag{6.3}$$

$$\text{s. t.} \quad B \geq 0.$$

The first term, $(\tilde{s} - t)(\tilde{D}_S - B)^+$, of the right hand side in (6.3) is the expected opportunity cost, the second term $g\tilde{Q}(B)$ represents the expected offload cost. Since the expected opportunity cost decreases in B (see below) and the expected offload cost increases, minimizing Γ implies finding the optimal cost trade-off.

Theorem 6.1. *If $g \geq \frac{\mathcal{E}[\tilde{s}] - t}{1 - G(x)}$, the optimal booking level B^* is the candidate solution that minimizes total expected overbooking cost Γ with*

---

[4] So far, the booking level has been assumed to be $B = K - N$.

[5] One may argue, though, that the offload cost increases in the offloaded quantity as loss-of-goodwill costs increase more than proportionally.

$$\text{argmin}[\Gamma(B)] \in \left\{ K - N, \underbrace{K - F_C^{-1}\left(1 - \frac{\mathcal{E}[\tilde{s}] - t}{g(1 - G(x))}\right)}_{B_0}, K \right\}. \quad (6.4)$$

Else, B* approaches infinity.

Proof. Differentiating (6.3) with respect to B gives

$$\frac{\partial \Gamma(B)}{\partial B} = -(\mathcal{E}[\tilde{s}] - t)[1 - (F_S(B))] + g\frac{\partial \mathcal{E}[\tilde{Q}]}{\partial B}. \quad (6.5)$$

With $\partial \mathcal{E}[\tilde{Q}]/\partial B$ from (B.16) and (B.18), respectively, this results in:

Case 1 $K - N < B < K$

$$\frac{\partial \Gamma(B)}{\partial B} = [1 - F_S(B)]\Big[g[1 - G(x)][1 - F_C(K - B)]$$
$$- (\mathcal{E}[\tilde{s}] - t)\Big] \stackrel{!}{=} 0. \quad (6.6)$$

If $g \geq \frac{\mathcal{E}[\tilde{s}]-t}{1-G(x)}$, (6.6) can be solved for B yielding

$$B_0 = K - F_C^{-1}\left(1 - \frac{\mathcal{E}[s] - t}{g(1 - G(x))}\right). \quad (6.7)$$

If B_0 falls within the limits of the case considered at present, i.e., $K - N < B_0 < K$, B_0 is a candidate solution. If $B_0 \leq K - N$, the candidate solution is $K - N$.

Case 2 $B > K$

The partial derivative with respect to B is:

$$\frac{\partial \Gamma(B)}{\partial B} = [1 - F_S(B)][gG(x) - (\mathcal{E}[\tilde{s}] - t)] \stackrel{!}{=} 0. \quad (6.8)$$

Since the null of this term (when $gG(x) = (\mathcal{E}[\tilde{s}] - t)$) is not a function of B, it follows that no optimum exists beyond K.

Case 3 $B = K$

Since $\tilde{Q}(B)$ and thus $\Gamma(B)$ are not differentiable at $B = K$, $B = K$ has to be considered as a candidate solution.

If $g < \frac{\mathcal{E}[\tilde{s}]-t}{1-G(x)}$, then $\partial \Gamma(B)/\partial B < 0$ for all values of B, i.e., total cost declines in B, and the optimal solution is choosing B as large as possible ($B^* \to \infty$).

$\square$

The expected offload quantity $\mathcal{E}[\tilde{Q}]$ cannot be determined numerically (see Sec. B.2). For the results presented in the following, the expected offload quantity therefore is approximated by simulation. Note that B_0 *can* be determined analytically from (6.7).

6.3.4 Illustration of Optimal Overbooking Policy

As pointed out above (see Sec. 6.3.2), overbooking is most relevant if spot market demand is sufficiently high such that denied spot market bookings are likely to occur and if spot market demand and price uncertainty are high. The optimal overbooking policy of the asset provider is therefore illustrated by the means of a scenario that exhibits these characteristics. Table 6.2 lists the values of those exogenous variables insofar as they deviate from the previously introduced base case.

Since one gets $g = 20 > \frac{\mathcal{E}[\tilde{s}]-t}{1-G(x)} = 15.82$, the optimal overbooking level B^* is determined according to Theorem 6.1 by first calculating the candidate solutions and then selecting the candidate solution that yields the lowest expected total cost. Table 6.3 lists the candidate solutions and the respective levels of expected total cost $\Gamma(B)$ in the high-uncertainty scenario. Here, it is $B^* = B_0 = 321.70$.

This is also illustrated in Fig. 6.6. The total expected cost curve $\Gamma(B)$ is the sum of the expected offload cost $g\mathcal{E}[\tilde{Q}]$ and the expected opportunity

Table 6.2. Values for exogenous variables in the high-uncertainty scenario

Variable[a]	Value	Description
$\sigma_{\tilde{D}_C}$	$\frac{1}{3}\mu_{\tilde{D}_C}$	Standard deviation of contract market demand
$\mu_{\tilde{D}_S}$	250	Mean of spot market demand
$\sigma_{\tilde{D}_S}$	$\frac{1}{3}\mu_{\tilde{D}_S}$	Standard deviation of spot market demand
$\sigma_{\tilde{s}}$	$\frac{1}{3}\mu_{\tilde{s}}$	Standard deviation of spot price
g	20	Offload cost per unit

[a] All other variables as in the base case (see Table 5.1).

Table 6.3. Candidate solutions for B^* in the high-uncertainty scenario

Candidate solution		Expected total cost
B	Value	$\Gamma(B)$
$K - N$	304.61	192.49
B_0	321.70	186.30[a]
K	400.00	204.25[a,b]

[a] Simulation result
[b] Left-hand limit for $B \to K$

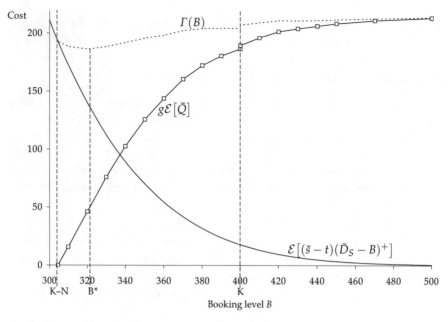

Fig. 6.6. Determination of optimal booking level B^*. The diagram shows the progression of the total expected cost curve $\Gamma(B)$, expected offload cost $g\mathcal{E}[\tilde{Q}]$ with $g = 20$, and expected opportunity cost $\mathcal{E}[(\tilde{s}-t)(\tilde{D}_S - B)^+]$.

cost curve $\mathcal{E}[(\tilde{s}-t)(\tilde{D}_S - B)^+]$. The expected offload cost and thus total expected cost are not continuous in B at the point $B = K - N$. This results from the form of $\mathcal{E}[\tilde{Q}]$ as in (B.11) and ultimately is a consequence of the not continuously differentiable form of $\tilde{E}$ in Lemma 4.1. The total expected cost curve has a global minimum at $B = 321.70$.

The shape of the total expected overbooking cost curve $\Gamma(B)$ depends on the per-unit offload cost g. Fig. 6.7 depicts the total expected overbooking cost for different levels of g. For low levels of g ($g = 10$ and $g = 12$) it can be seen that the total expected cost is always decreasing in B. Thus, the asset provider would overbook capacity without limit ($B \to \infty$) as the expected loss from overbooking is always smaller than the expected benefit. Vice versa, for high levels of g ($g = 26$ and beyond), the total expected cost is always increasing in B. Thus, the asset provider would not overbook capacity at all ($B = K - N$) because offloading is too expensive. For the levels of g in-between, the total expected cost curve exhibits a minimum between $B = K - N$ and $B = K$ that represents the asset provider's optimal booking level.

Analytically, the threshold levels of g can be shown to be

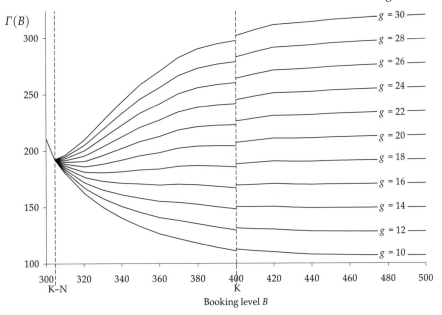

Fig. 6.7. Progression of total expected overbooking cost curve $\Gamma(B)$ for different levels of offload cost g: For low levels of g, $\Gamma(B)$ is convex, for high levels of g, $\Gamma(B)$ is concave in B.

$$\underline{g} = \frac{\mathcal{E}[\tilde{s}] - t}{1 - G(x)} \tag{6.9a}$$

and

$$\overline{g} = \frac{\mathcal{E}[(\tilde{s} - t)(\tilde{s} - x)^{+}]}{r[1 - G(x)]} \tag{6.9b}$$

where

$$B^{*} \to \infty \qquad \text{if } g \leq \underline{g} \tag{6.10a}$$

$$B^{*} = B_{0} \qquad \text{if } \underline{g} < g \leq \overline{g} \tag{6.10b}$$

$$B^{*} = K - N \quad \text{if } \overline{g} < g \tag{6.10c}$$

The lower threshold level in (6.9a) follows directly from the proof to Theorem 6.1. The upper threshold level in (6.9b) can be derived by substituting N from (4.12) and B from (6.7) into the condition $K - N < B < K$ in the proof of Theorem 6.1 and solving for g.

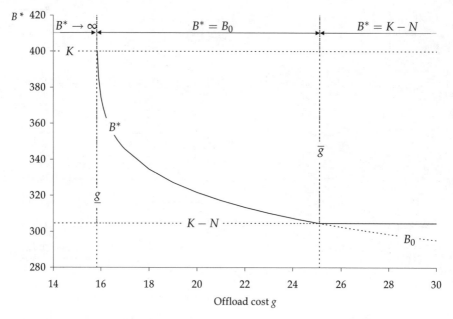

Fig. 6.8. Optimal booking level B^* as a function of the per-unit offload cost g in the high-uncertainty scenario.

With this, the optimal booking level B^* can be depicted as a function of the per-unit offload cost g (see Fig. 6.8). In the high-uncertainty scenario, it is $\underline{g} = 15.82$ and $\overline{g} = 25.10$.

Though it may be difficult to assess g in reality, the order of magnitude of g may very well fall within this interval. The expected margin (with $\mathcal{E}[\tilde{s}] - t = 15$) establishes presumably the lowest possible level of g, if a customer can be turned down without harming goodwill or requiring any compensation. If the customer is compensated by free service at another occasion, the offload cost amounts to roughly the expected spot price (with $\mathcal{E}[\tilde{s}] = 20$). Only if the asset provider needed to pay further compensation or incurred additional loss-of-goodwill cost, g would eventually exceed $\overline{g}$.

Expressed in percentage terms, the asset provider optimally overbooks the spot market allotment $K - N$ by 31% at maximum for $g = \underline{g}$ in the high-uncertainty scenario, increasing the asset provider's expected profit by approximately 2%.

6.3.5 Joint Optimization of Tariff and Booking Level

So far, the optimal booking level B^* has been derived by taking the optimal tariff (r^*, x^*) as given. However, since the tariff (r^*, x^*) implicitly given by (4.23) and (4.24) is a function of the booking level, which had been assumed

Table 6.4. Iterations for determining r^{**}, x^{**}, and B^{**} in the high-uncertainty scenario

i	0	1	2	3	4	5 ...	33	34
r_i^*		6.920	6.515	6.832	6.581	6.778 ...	6.690	6.690
x_i^*		9.169	9.261	9.189	9.245	9.201 ...	9.221	9.221
N_i^*		95.387	100.698	96.534	99.825	97.241 ...	98.384	98.387
B_i		304.613	316.392	307.105	314.407	308.654 ...	311.180	311.186
B_i^*		321.703	320.031	321.337	320.302	321.113 ...	320.753	320.752
ΔB_i	0	17.090	3.639	14.232	5.895	12.460 ...	9.573	9.566

to be $B = K - N$ (no overbooking), the tariff (r^*, x^*) is no longer optimal if $B > K - N$, i.e., in the case of overbooking.

Finding an optimal solution that takes this interdependency into account requires joint optimization of the asset provider's objective function with overbooking:

$$\max_{r,x,B} \Pi_A - g\mathcal{E}[Q] \tag{6.11}$$

$$\text{s. t.} \quad r \geq 0, \ x \geq 0, \ B \geq 0$$

Let (r^{**}, x^{**}, B^{**}) denote the solution to the joint optimization problem. Unfortunately, solving the problem directly for (r^{**}, x^{**}, B^{**}) is infeasible. It has been shown in Chap. 4 that the determination of the optimal tariff r^*, x^* requires a numerical solution procedure which in turn presupposes a formulation of the expected value of the objective function that can be submitted to an optimization software. In Sec. 6.3.3 it was noted that the expected value of the objective function with overbooking cannot be formulated in near-closed form because of the expression for the expected offload quantity (see Sec. B.2). The objective function with overbooking thus cannot be optimized via numerical procedures by an optimization software. Because of this incompatibility of the solution approaches, the solution to the joint optimization problem in (6.11) needs to be approximated by the following iterative algorithm:

Step 0 (Initialize) Set as starting value

$$B_0 = K - N_0. \tag{6.12}$$

Step 1 (Maximize profit) Let i denote the number of the current iteration, starting at $i = 1$. Determine r_i^* and x_i^* according to Theorem 4.2 with $N_i^* = N_i^*(r_i^*, x_i^*)$ according to Theorem 4.1 by calculating

$$\max_{r_i, x_i} \Pi_A \tag{6.13}$$

$$\text{with}\quad B_i = K - N_i^* + \Delta B_{i-1}\quad\text{for}\quad i > 0.$$

Step 2 (Minimize overbooking cost) With the results from Step 1, determine B_i^* according to Theorem 6.1 by calculating

$$\min_{B_i} \Gamma(r_i^*, x_i^*) \tag{6.14}$$

and let

$$\Delta B_i = B_i^* - B_i. \tag{6.15}$$

Step 3 (Convergence test) Let the stopping criterion $\delta > 0$ be a small number. If $|r_i^* - r_{i-1}^*| < \delta$ and $|x_i^* - r_{i-1}^*| < \delta$ and $|B_i^* - r_{i-1}^*| < \delta$, then stop and take $r_i^* \approx r^{**}$, $x_i^* \approx x^{**}$, and $B_i^* \approx B^{**}$ as the solution to the joint optimization problem in (6.11); else continue at Step 1 with $i \leftarrow i + 1$.

This algorithm is applied in the following to determine the jointly optimal solution in the high-uncertainty scenario. Table 6.4 shows the results of the first five and last two iterations. After 34 iterations, the desired precision of $\delta = 10^{-3}$ is reached. Fig. 6.9 illustrates exemplarily for the three decision variables the evolution of B_i^* over the course of the iterations. B_i^* oscillates around and converges against $B^{**} = B_{34}^* = 320.75$.

The results of the three optimization procedures for the asset provider's expected profit are summarized in Table 6.5: The result without overbooking corresponds to the solution in the high-uncertainty scenario according to Theorem 4.2, the result of the sequential optimization according to Theorem 6.1 taking the result from Theorem 4.2 as given, and the result of the joint optimization by applying the above iterative algorithm.

Table 6.5 shows that the original solution using sequential optimization results in setting the booking level B and the reservation fee r slightly too

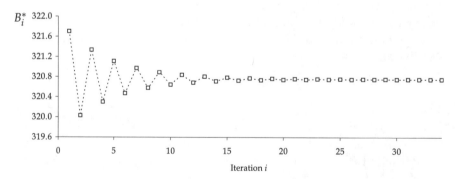

Fig. 6.9. When the number of iterations i increases, B_i^* converges against B^{**}

Table 6.5. Comparison of expected profits in the high-uncertainty scenario

	Booking level B	Tariff r x	Reservations N^*	Expected incremental profit[a]
Without overbooking	304.61	6.92 9.17	95.39	0
With overbooking				
Sequential optimization	321.70	6.92 9.17	95.39	5.62[b]
Joint optimization	320.75	6.70 9.22	98.32	7.70[b]

[a] In comparison to situation without overbooking
[b] Simulation result

high and the execution fee at slightly too low a level. This can be best explained by first considering the execution fee.

The higher the execution fee, the higher *ceteris paribus* the probability that $\tilde{s} < x$ and thus the probability of options not being executed. In the optimization without overbooking, non-execution of options always means an opportunity loss to the asset provider since the capacity cannot be sold elsewhere. In the joint optimization with overbooking, the capacity on which the asset provider has written and sold options can at least partially be sold elsewhere, namely in the spot market. It is thus rational that the execution fee in the optimization without overbooking is lower than in the joint optimization with overbooking. In turn, the asset provider can lower the reservation fee and thus trigger a higher number of reservations.

With regard to the reaction of the optimal booking level to tariff changes, there are two opposite effects. On the one hand, one would expect the optimal booking level to increase when the execution fee increases because – as pointed out above – the expected number of non-executed options increases and the asset provider then enlarges the provision for non-executed options by increasing the amount of overbooked capacity. On the other hand, an increase in the reservation or execution fee lets the expected contract market demand decrease (due to the price sensitivity) and thus also adversely affects the optimal overbooking level.

The total effect in the case of the high-uncertainty scenario is negative, i.e., the increase of x and N and the decrease of r lead to a slight decrease of the booking level B in absolute terms as compared to the sequential optimization approach. The relative amount of overbooking, measured as $[B - (K - N)]/(K - N) - 1$, however, increases when applying joint optimization (relative overbooking amounts to 6.3%) instead of sequential optimization (5.4%) and confirms the original expectation about the reaction of the optimal booking level to tariff changes.

Table 6.5 informs also about the goodness of the sequential optimization solution by indicating the asset provider's expected profits. The full profit improvement potential of overbooking is realized by the joint optimization approach. The sequential optimization approaches realizes 73% of the full total profit improvement potential in the high-uncertainty scenario.

Application Case Study in the Air Cargo Industry

As demonstrated in Chap. 5, the application of capacity-option contracts can generally result in an increase of the market participants' financial performance under the contract. It remains to be investigated, though, if and to what extent the actual specification of the market characteristics in the air cargo industry favor the realization of these benefits. As a first attempt into this direction, this chapter reports on an application case study conducted on the basis of a data set provided by Lufthansa Cargo AG (see also Sec. 2.2). The following sections introduce the data sample and describe the estimation of the model parameters from the sample. The estimates are then submitted to the model for analysis, followed by a presentation of the resulting optimal pricing policies and incremental revenues expected to be earned by asset provider and intermediary.

7.1 Data Sample

The data sample was provided by Lufthansa Cargo AG and covers the time period of the carrier's winter flight schedule 2003/2004. For each considered flight, data from three different sources were combined, including data on the flight's total utilization (see Table 7.1 for a sample of these data), standing contracts (Table 7.2), and attained average yield (Table 7.3).

A *flight* designates in the following time-series data for a specific flight number on a specific day of the week over a period of multiple (13–26) weeks, i.e., flight no. LH1234 on day 3 (Wednesday) and day 6 (Saturday) are considered two different flights. In total, twelve flights were considered for analysis, of which six could finally be included. The remainder was sorted out due to insufficient data availability and/or consistency between the three data sources. The original data set contained data on seven cargo-only and five mixed passenger-cargo flights, the analyzed set comprises four

Table 7.1. Example for utilization data of a particular flight (disguised)

Flight no.	Date	Day of week	Itinerary Origin	Destination	Weight (kg) Demand	Supply	Volume (m^3) Demand	Supply
LH1234	08.11.03	6	FRA	JNB	24,392	45,000	247	257
LH1234	15.11.03	6	FRA	JNB	42,277	45,000	255	257
LH1234	22.11.03	6	FRA	JNB	35,757	45,000	227	257
⋮	⋮	⋮	⋮	⋮	⋮	⋮	⋮	⋮

Table 7.2. Example for allotment data of a particular flight (disguised)

Flight no.	Date	Day of week	Allotment ID	Allotment weight (kg) Contract	Used	Allotment volume (m^3) Contract	Used	Rate (€/kg)
LH1234	08.11.03	6	FRAxyJNBIF	1,800	1,800	10	10	1.48
LH1234	15.11.03	6	FRAxyJNBIF	1,800	1,208	10	8	1.48
⋮	⋮	⋮	⋮	⋮	⋮	⋮	⋮	⋮
LH1234	08.11.03	6	FRAxyJNBAL	7,500	7,315	45	44	1.42
LH1234	15.11.03	6	FRAxyJNBAL	7,500	7,500	45	45	1.42
⋮	⋮	⋮	⋮	⋮	⋮	⋮	⋮	⋮

Table 7.3. Example for yield data of a particular flight (disguised)

Flight no.	Date	Day of week	Leg Depature	Arrival	Yield (€/kg)
LH1234	08.11.03	6	FRA	NBO	1.76
LH1234	15.11.03	6	FRA	NBO	1.91
LH1234	22.11.03	6	FRA	NBO	1.85
⋮	⋮	⋮	⋮	⋮	⋮

cargo-only flights and two mixed passenger-cargo flights. To protect confidentiality, data and results are reported on in the following in aggregated (indexed) or disguised form only.

7.2 Estimation of Model Variables

The model input variables, including foremost capacity, contract market demand, spot market demand and price, are estimated from the above data set, separately for each flight.

7.2.1 Capacity

For cargo-only flights, capacity is taken directly from the utilization data. For some cargo-only flights, capacity was split-up between the considered carrier and an airline-alliance partner. In these cases, only the carrier's own capacity share entered the analysis (the same applies for contract, demand, and yield data). For mixed passenger-cargo flights, the utilization data only showed the originally planned and for-certain available cargo capacity. Since actual cargo volume for almost all data points exceeded this capacity, the total capacity variable K was estimated to result in an average capacity utilization at the lower end of the capacity utilization range of the cargo-only flights in the sample (see also Sec. 2.2.3.1 for average past utilization data).

In general, the unit of capacity used in the following is kilogram, i.e., capacity is specified in terms of weight and not of volume. The assumption underlying this simplification is that on average the specific weight, i.e., weight per volume, is constant. Weight was preferred to volume as capacity unit because rates are usually weight-based (however, surcharges in form of the so called "chargeable weight" for highly voluminous freight apply).

7.2.2 Contract Market Demand

As defined in Sec. 4.2, the capacity-option model treats contract market demand as endogenous, derived from the contract market demand function which is given by the exogenous parameters a and b as well as the endogenous price p. Since parameters a and b are hard to observe in reality, the approach taken here first estimates the parameters $\mu_{\bar{D}_C}$ and $\sigma_{\bar{D}_S}$ of the contract market demand distribution and then draws conclusions about a and b.

7.2.2.1 Parameters of the Demand Distribution

The data set contains information on contract market demand in the form of the called-on reservations. This information, however, displays per se not the entire demand distribution but is truncated at the reservation limit.[1] At-

---

[1] This does not hold strictly for the entire data set, because the number of called-on reservations sporadically may exceed the number of reservations, if the carrier allowed its customer to overdraw his allotment, while in the same data series other observations are curtailed at the reservation level.

taining estimates of mean and standard deviation of the non-truncated underlying distribution requires unconstraining of the observed data (cf. Orkin 1998; Weatherford and Pölt 2002; Talluri and van Ryzin 2004, p. 473 ff.). To this end, the following approach is taken for each standing long-term contract:

E_j denotes the capacity called on at observation j, N the reserved amount of capacity, and J the total number of reservations. There are C observations censored at $E_j = N \equiv E_j^{(0)}$ with $j = J - C + 1, \ldots, J$ and $C < J$; all other observations are uncensored with $j = 1, \ldots, J - C$. Unconstrained estimates of mean μ and standard deviation σ of the underlying demand distribution can be attained by the following iterative Expectation-Maximization algorithm (based on Talluri and van Ryzin 2004, pp. 476–478; Talluri and van Ryzin 2005, p. 2):

Step 0 (Initialize) Set $\mu^{(0)}$ and $\sigma^{(0)}$ as starting values for μ and σ, using all available observations.

$$\mu^{(0)} = \frac{\sum_{j=1}^{J} E_j}{J} \tag{7.1}$$

$$\sigma^{(0)} = \sqrt{\frac{\sum_{j=1}^{J} \left(E_j - \mu^{(0)}\right)^2}{J - 1}} \tag{7.2}$$

Step 1 (Replace censored data) Let i denote the number of the current iteration, starting at $i = 1$. For all censored data, i.e., for $j = J - C + 1, \ldots, J$, calculate:

$$E_j^{(i)} = E_j^{(0)} + \sigma^{(i-1)} \varphi \left(z_N^{(i-1)}\right) - \left[N - \mu^{(i-1)}\right] \left[1 - \Phi \left(z_N^{(i-1)}\right)\right] \tag{7.3}$$

$$\text{where} \quad z_N^{(i-1)} = \frac{N - \mu^{(i-1)}}{\sigma^{(i-1)}} \tag{7.4}$$

Step 2 (Recalculate moments) The new estimates of μ and σ are:

$$\mu^{(i)} = \frac{\sum_{j=1}^{J-C} E_j + \sum_{j=J-C+1}^{J} E_j^{(i)}}{J} \tag{7.5}$$

$$\sigma^{(i)} = \sqrt{\frac{\sum_{j=1}^{J-C} \left(E_j - \mu^{(i)}\right)^2 + \sum_{j=J-C+1}^{J} \left(E_j^{(i)} - \mu^{(i)}\right)^2}{J - 1}} \tag{7.6}$$

Step 3 (Convergence test) Let the stopping criterion $\delta > 0$ be a small number. If $\left|\mu^{(i)} - \mu^{(i-1)}\right| < \delta$ and $\left|\sigma^{(i)} - \sigma^{(i-1)}\right| < \delta$, then stop and let $\mu \approx \mu^{(i)}$ and $\sigma \approx \sigma^{(i)}$; else continue at Step 1 with $i \leftarrow i + 1$.

For each date in the data series, the unconstrained demand observations (with $\delta = 10^{-3}$) of all contract holders are added up, yielding total demand from contract holders. The average total demand from contract holders and the standard deviation thereof are taken as estimators for the mean $\mu_{\tilde{D}_C}$ and standard deviation $\sigma_{\tilde{D}_C}$ of contract market demand (assuming that individual contract demands are mutually independent).

7.2.2.2 Parameters of the Demand Function

Under the assumption that (1) demand follows – at least for the considered region of the demand function – a linear demand curve, that (2) the combination of average contract market demand $\mu_{\tilde{D}_C}$ and average contract rate w_0 represents one point of the demand curve, and (3) given the price elasticity of demand η at this point as defined in (5.10),[2] the parameters of the linear demand function can be determined by solving the set of equations:

$$a = \mu_{\tilde{D}_C} + bp_0 \tag{7.7}$$

$$b = -\frac{a\eta}{p_0(1 - \eta)} \tag{7.8}$$

where $p_0 = w_0 + \lambda$ (for λ, see Sec. 7.2.5). w_0 is the average contract rate of all capacity agreements on that particular flight, weighted by the individually reserved amount of capacity. Individual contract rates differ by customer and contract type (with or without return clause, i.e., GCA or CPA, see Sec. 2.2.3). Since for the latter no systematic rate difference could be identified, i.e., no contract type exhibited a systematic higher rate than the other, the average was taken over all existing contracts.

7.2.3 Spot Market Demand

Spot market sales of the asset provider were estimated as the difference of total capacity sold (from utilization data) and called-on capacity (from allotment data). Comparing spot market sales with available capacity in the spot market (total capacity supply from utilization data less called-on capacity) did not indicate necessity for unconstraining. In principle, the above unconstraining algorithm could be applied if necessary. Mean $\mu_{\tilde{D}_S}$ and standard deviation $\sigma_{\tilde{D}_S}$ of spot market demand were here estimated directly from time series of spot market sales.

---

[2] Since the data set allows to draw only limited conclusions about the actual price elasticity of demand, the results presented below are reported over a range of η.

7.2.4 Spot Price

The yield attained in the spot market is not recorded directly by the carrier who provided the data sample. However, the carrier tracks the average yield generated on the first leg of each flight. A leg designates a non-stop segment of a flight. Non-stop (direct) flights are thus single-leg flights, flights with stopovers are multi-leg flights. For single-leg flights, the spot market yield for each observation was calculated by first calculating spot market revenue as the difference of total revenue (average total yield from yield data multiplied by total demand from utilization data) and contract market revenue (called-on reservations multiplied by the respective contract rate). Dividing the resulting spot market revenue by the capacity sold in the spot market (see Sec. 7.2.3) gives the spot market yield for each observation. The expected spot price $\mu_{\tilde{s}}$ (arithmetic mean) and its standard deviation $\sigma_{\tilde{s}}$ were estimated from time series of spot market yields.

For multi-leg flights, the attained yield of the first leg (from yield data) had to be projected to reflect the yield of the entire flight. Because no other yield data was available, it was assumed that the ratio of total-flight yield to first-leg yield can be approximated by the ratio of the carrier's published standard price for a standard spot market product from the flight's origin to its final destination[3] to the published price from the flight's origin to its first stopover. In addition, it was assumed that two thirds[4] of all cargo travel to the final destination.

7.2.5 Other Parameters

The intermediary's markup on the carrier's contract rate was assumed to be 50%, i.e., $\lambda = 0.5 \times w_0$; this assumption has no influence on the structure of the results presented below. Furthermore, all variable cost parameters were assumed to equal zero. Discussions with the carrier who provided the data sample revealed that the carrier does not assign any specific variable cost to the usage of its capacity, but considers all expenses for a particular scheduled flight (including cost for capacity, fuel, and handling) as quasi-fixed. Fixed cost, which have no influence on the model's solution procedure anyhow, were likewise assumed to equal zero. The carrier's optimization objective thus shifts from profit to revenue maximization.

7.3 Results

Following the above procedure, the model parameters were estimated for the six sample flights. Table 7.4 gives an overview of the estimates and the

---

[3] All capacity agreements in the data sample referred to this relation.

[4] Simplification, based on private communication with the data-providing carrier.

Table 7.4. Model parameters estimated from data sample

Flight no.[a]	Day of week	Number of contracts[b]	Total reserved capacity N_0	Implied service level[c]	Average contract rate w_0 (indexed)	Capacity K (indexed)	$\mu_{\tilde{D}_C}$	$\sigma_{\tilde{D}_C}$	$\mu_{\tilde{D}_S}$	$\sigma_{\tilde{D}_S}$	$\mu_{\tilde{s}}$	$\sigma_{\tilde{s}}$	Total expected demand[d]
M1	3	2 (0)	405.6	99%	1.00	1,000	258.7	58.4	423.9	58.8	2.46	0.99	682.6
M1	6	3 (0)	588.9	97%	1.00	1,000	259.9	172.4	544.0	114.5	2.19	0.72	803.9
C1	7	7 (7)	457.0	57%	1.00	1,000	430.5	140.9	499.3	98.7	1.24	0.21	929.9
C2	4	3 (1)	149.1	78%	1.00	1,000	117.4	41.5	754.4	136.3	1.38	0.13	871.8
C3	6	1 (1)	40.3	34%	1.00	1,000	49.1	21.9	797.6	97.5	1.37	0.17	846.7
C3	7	6 (2)	521.1	69%	1.00	1,000	455.9	132.4	346.7	147.2	2.26	0.77	802.6

[a] M indicates mixed passenger-cargo flight, C indicates cargo-only flight.
[b] In parentheses: thereof contracts without return clause (CPA).
[c] Calculated as $(N_0 - \mu_{\tilde{D}_C})/\sigma_{\tilde{D}_C}$ (see p. 62).
[d] Calculated as $\mu_{\tilde{D}_C} + \mu_{\tilde{D}_S}$.

standing capacity agreements for each flight. All quantity data is reported on relative to total capacity K, which is indexed at $K = 1,000$ both to allow for better comparability and to protect confidentiality; likewise, all pricing data is reported on relative to the average actual contract rate w_0, which is indexed at $w_0 = 1$.

When willing to accept – as an unintended side effect of the normal distribution assumption – at most a probability of 1% (equivalent to a co-efficient of variation $\vartheta = 0.43$) of having negative quantities or prices that reduce the model's accuracy, flight M1 on Saturdays exhibits too high a contract market demand uncertainty ($\vartheta_{\tilde{D}_C} = 0.66$) to be submitted to the model for analysis. This is due to the fact that reserved capacity was only called-on sporadically for this particular flight, while reserved capacity exceeded average contract market demand by more than the twofold. Flight M1 on Saturdays is thus excluded from the following analysis.

For the remaining five flights, the optimal reservation and pricing policies were determined by applying the model introduced in Chap. 4. Since the data sample did not provide sufficient information to estimate the price responsiveness of contract market demand, the analysis was conducted over a range of price elasticities of contract market demand η (see Sec. 7.2.2.2).

In general, the analysis reveals that the actual number of capacity reservations and the average contract rate w_0 in the data set deviate from the optimal values as derived by the model. Assuming that the actual number of capacity reservations (see Table 7.4) were optimal, the corresponding price w_0 would be too low or, assuming that w_0 were the optimal price, the number of capacity reservations would be too high. This mismatch can be explained by the fact that the model assumes that all contracts can be enforced while the contracts in the sample data were – even in those cases where the terms of the contract did not include a return clause – not strictly enforced. One can thus conclude that forwarding companies anticipated this non-enforcement and systematically overstated their actual capacity requirements, implying in the most extreme case a service level of up to 99%.

Furthermore, one can observe in Table 7.4 that the characteristics of the flights in the sample differs widely, both in terms of the share of total capacity that is sold in the contract market, the amount of demand in the contract market, and the yield attainable in the spot market. One reason for this heterogeneity is the fact that the flights serve different geographic markets. However, even flights to the same destination – in the sample, e.g., flight C3 – may exhibit different characteristics, which is due to the flights' day of the week. Flight C3 on Sundays has much stronger demand in the contract market than on Saturdays and yields a relatively higher expected spot price (see Sec. 2.2.3.3).

7.3.1 Optimal Choice of Contract Type and Pricing Policy

For the five analyzed flights, Fig. 7.1 shows the optimal pricing policies and incremental revenues as a function of the absolute price elasticity of contract market demand $|\eta|$[5]. The left-hand side diagrams in Fig. 7.1 show the optimal pricing policies. Relative to the average actual contract rate $w_0 = 1$, $\mu_{\tilde{s}}$ indicates the estimated average spot price. One can notice that for two flights, in Fig. 7.1(a) and (i), respectively, the expected spot price considerably (more than twofold) exceeds the average actual contract rate w_0, while for the other flights $\mu_{\tilde{s}}$ is less than 50% greater than w_0.

Likewise relative to w_0, w^* indicates the optimal capacity price of the (credibly enforceable) fixed-commitment contract from the capacity-option pricing model. Interestingly, one can observe that for all flights w^* is indeed around the carrier's actual price w_0, for some flights, see in particular Fig. 7.1(c) and (i), even very close to w_0 (at least for some ranges of $|\eta|$, as discussed below). This can be seen as a confirmation of the model's ability to capture the essentials of the dynamics of spot and contract market in a

---

[5] Because it has been assumed in Sec. 4.2.2 that $b \geq 0$ and thus the slope of the demand function $-b \leq 0$, η is a negative number. By convention (see Sec. 5.4.2.3), "high" and "low" with regard to price elasticity here always refer to the *absolute* value of η, i.e., $|\eta|$. For $|\eta| < 1$, demand is referred to as (price-) inelastic; for $|\eta| \geq 1$, as (price-) elastic.

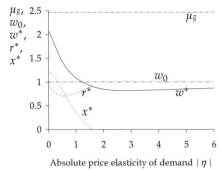

(a) Flight M1 on Wednesdays: Prices (indexed)

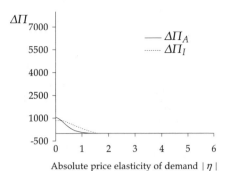

(b) Flight M1 on Wednesdays: incremental revenues

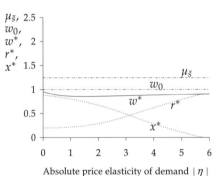

(c) Flight C1 on Sundays: Prices (indexed)

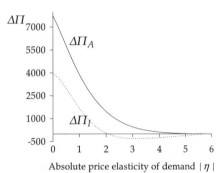

(d) Flight C1 on Sundays: incremental revenues

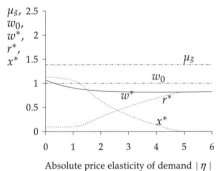

(e) Flight C2 on Thursdays: Prices (indexed)

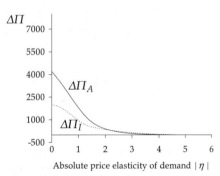

(f) Flight C2 on Thursdays: incremental revenues

Fig. 7.1. Optimal pricing policies (indexed) and incremental revenues for sample flights

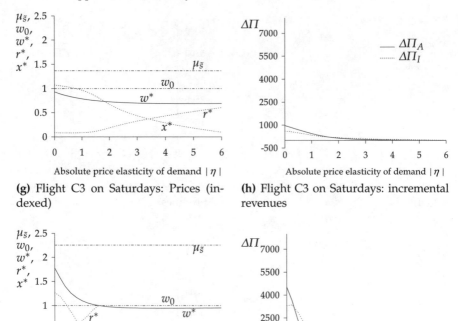

(g) Flight C3 on Saturdays: Prices (indexed)

(h) Flight C3 on Saturdays: incremental revenues

(i) Flight C3 on Sundays: Prices (indexed)

(j) Flight C3 on Sundays: incremental revenues

Fig. 7.1. Optimal pricing policies (indexed) and incremental revenues for sample flights (cont.)

similar fashion as the carrier's pricing department takes those into account when setting capacity prices in long-term capacity contracts, especially since w_0 is not an input parameter of the capacity-option pricing model, but was only used as a reference point in the estimation of the contract-market demand curve.

As a consequence of the presence of contracts with return clauses and of forwarders anticipating the non-enforcement of contracts without return clauses, w^* is, at least for higher values of $|\eta|$, lower than w_0. Only in the price-inelastic region (and slightly beyond, up to $|\eta| \approx 1.5$) of the functions in Fig. 7.1(a) and (i), w^* considerably exceeds w_0. However, conjecturing that the price w_0 set by the carrier is in the order of magnitude of the optimal price, the actual value of w_0 can be interpreted as an indicator that forwarder's demand is in reality price-elastic for these flights.

In addition to the fixed-commitment tariff, the diagrams at the left-hand side of Fig. 7.1 also indicate the optimal pricing policy for a capacity-option contract with reservation fee r^* and execution fee x^* for those regions of $|\eta|$ where a capacity-option contract yields higher expected profit for the asset provider than a fixed-commitment contract. In Fig. 7.1(a) and (i), the region of $|\eta|$, where the capacity-option contract is optimal, is restricted to the price-inelastic region of $|\eta|$ and a small part of the price-elastic region (up to $|\eta| \approx 1.5$). Because of the above conclusion that actual price elasticity is presumably in the price-elastic region of $|\eta|$, it is probably optimal to offer only fixed-commitment contracts on flights M1 on Wednesdays and C3 on Sundays, which are the flights exhibiting a high expected spot market yield relative to the actual contract rate w_0 (see above). For flights C1 on Sundays, C2 on Thursdays, and C3 on Saturdays, capacity-option contracts turn out to be the optimal contract choice for a wide range of price elasticity of demand (up to $|\eta| \approx 5.8$, $|\eta| \approx 5.0$, and $|\eta| \approx 7.4$, respectively). On these flights, the average contract rate is closer to the expected spot market yield than for the other two.

7.3.2 Incremental Revenue Earned by Using Capacity-Option Contracts

The diagrams on the right-hand side of Fig. 7.1 show the incremental expected revenues for the asset provider ($\Delta\Pi_A$) and intermediary ($\Delta\Pi_I$). Incremental revenue refers to the additional revenue the two market participants can expect to earn when using capacity-option instead of fixed-commitment contracts whenever the choice of a capacity-option contract is optimal for the asset provider. Revenue increments are denoted in Euro and based on actual (not indexed) data.

In general, revenue increments decrease with absolute price elasticity of demand $|\eta|$, i.e., the more demand is price-elastic, the smaller the additional revenue the asset provider can generate by using capacity-option instead of fixed-commitment contracts. For high values of $|\eta|$ for which the optimal contract choice is a fixed-commitment contract, the revenue increment becomes zero for any flight (this corresponds to the findings in Sec. 5.4.2.3).

For all but one flight, the increase of expected revenue for the asset provider goes hand in hand with an increase of expected revenue for the intermediary. Only for flight C1 on Sundays, the intermediary suffers a decrease in expected revenue for the region of $|\eta|$ from ca. 2 to 5.8. This is due to the fact that on flight C1 the ratio of capacity to total demand is smaller than on the other considered flights (see the column labeled "Total expected demand" in Table 7.4), i.e., capacity is more scarce, and corresponds to the findings in Sec. 5.4.1.

Table 7.5. Studies on price elasticity of demand for air cargo transport

Study	Carrier type	Sample	$|\eta|$
Talley and Schwarz-Miller (1988)	Mixed passenger-cargo	USA, 22 carriers, 1983	1.32
Wang et al. (1981)	Mixed passenger-cargo	USA, 1950–1977	2.33 to 2.50
	Cargo-only		0.42 to 0.84
	Both types (aggregate model)		1.47 to 1.60
Oum et al. (1990)[a]	Both types	Unknown	0.82 to 1.60

[a] Survey article; elasticity range estimate includes Talley and Schwarz-Miller (1988), Wang et al. (1981), and a not further specified third study.

Two drivers are mainly influencing the order of magnitude of the achievable incremental revenues: the amount of expected demand in the contract market and the price elasticity of demand.

Considering the amount of expected demand in the contract market, it follows directly that the higher the demand in the contract market is, the higher is the lever of the optimal choice of the contract type used in the contract market. For flight C1 on Sundays in Fig. 7.1(d), e.g., with initially[6] $\mu_{\tilde{D}_C} = 430.5$, the incremental revenues would reach more than € 7,000 for the asset provider per flight event if price elasticity were very low. In contrast, for flight C3 on Saturdays in Fig. 7.1(h), though capacity-option contracts are better than fixed-commitment contracts for a wide range of $|\eta|$, the incremental revenues are rather small (less than € 1,000 for the asset provider at most) because contract market demand is rather low with initially $\mu_{\tilde{D}_C} = 49.1$.

With regard to the price elasticity of demand, Table 7.5 gives an overview of studies that estimate price elasticity of demand for air cargo transportation. Though differing in geographic scope and time period considered, these estimates can at least provide at an indication about actual price elasticity. The studies mostly report a range rather than a point estimate of price elasticities, resulting from the fact that price elasticities particulary differ on routes where capacity is particularly scarce in supply and those were capacity is amply available. Furthermore, a carrier can try to influence the price elasticity of the demand he himself faces, e.g., through product differentiation and customer relationship management. As described in Sec-

---

[6] Initially refers to the demand estimate on the basis of the average contract rate w_0 as indicated in Table 7.4. For contract rates different from w_0 – be it w^* or the split tariff (r^*, x^*) – demand differs from this value, following the demand function estimated in Sec. 7.2.2.2.

Table 7.6. Incremental expected revenue of asset provider ($\Delta\Pi_A$) and intermediary ($\Delta\Pi_I$)

Flight no.	Day of week		Absolute price elasticity of demand $|\eta|$					
			0.5	1	1.5	2	2.5	3
M1	3	$\Delta\Pi_A$	406	59	0	0	0	0
		$\Delta\Pi_I$	681	323	29	0	0	0
C1	7	$\Delta\Pi_A$	5,777	3,890	2,444	1457	830	453
		$\Delta\Pi_I$	2,995	1,666	650	52	−225	−305
C2	4	$\Delta\Pi_A$	3,052	1,801	836	394	191	90
		$\Delta\Pi_I$	1,638	955	527	357	233	147
C3	6	$\Delta\Pi_A$	697	445	253	143	86	54
		$\Delta\Pi_I$	492	342	233	176	135	104
C3	7	$\Delta\Pi_A$	1,930	338	1	0	0	0
		$\Delta\Pi_I$	2,633	1,229	83	0	0	0

tions 2.2.1 and 2.2.2, Lufthansa Cargo AG is undertaking efforts in both directions by introducing new products and service packages and establishing the Business Partnership Program. The studies on price elasticity agree that demand for freight transport tends to be less elastic than demand for passenger transport (cf. Oum et al. 1992; Doganis 2002, p. 307 ff.) since freight transport is a derived demand (see Sec. 2.1.2).

Table 7.6 lists the incremental revenues expected by asset provider and intermediary for values of price elasticity of demand selected on the basis of the estimates provided in Table 7.5. Taking the range of 0.5 to 2.5 for $|\eta|$ as an realistic estimate, the revenue increment amounts for some flights up to approximately € 5,800 for the asset provider and € 3,000 for the intermediary per flight event. With 26 flight events in a contract with a 6-month lifespan for a weekly flight, this is equivalent to a joint revenue increase of more than € 200,000 for flight C2 on Wednesdays.

These values represent rather modest estimates of the actual gains that might be achieved by using option contracts because the benchmark, which the performance of capacity-option contracts is measured against here, is the performance of a perfect, i.e., optimally priced, and enforceable fixed-commitment contract. Taking the mismatch between actual contract rate w_0 and total reserved capacity (as pointed out above) as an indication for the non-optimality of the currently employed contracts and conjecturing that capacity options facilitate contract enforcement (this aspect is discussed in Chap. 8), the performance increase resulting from capacity-option contracts is likely to be even greater if compared to the actual status quo.

The discussion of the managerial implications of this application case study is deferred to the concluding Chap. 8.

8

Managerial Implications and Conclusion

The major contribution of this thesis is the development of a capacity-option pricing model for long-term capacity reservation contracts in the air cargo industry. It has been shown that capacity-option contracts provide a way to price flexibility, allowing the buyer to react to movements of stochastic demand and prices and at the same time remunerating the seller for holding capacity ready for usage by the buyer. By the implementation of a split-tariff scheme, composed of a reservation fee payable ex ante and an execution fee payable only for those capacity units actually called on, capacity-utilization, demand, and price risk can be effectively shared between buyer and seller, leading to a more efficient market outcome. Through risk sharing among the contract parties, the performance of the capacity supply chain is increased and brought closer to the theoretic optimum of a supply chain governed by a single decision-maker (integrated firm). Because it is "the size of the pie" that is grown this way, it is possible – in most cases – to create a win-win situation for capacity buyer and seller, i.e., achieve a Pareto improvement.

In Chap. 5, it has been discussed which factors influence the very existence and the size of the Pareto improvement. It has been shown that under certain circumstances a Pareto improvement may not arise per se, making the intermediary potentially worse off when buying capacity options instead of signing a fixed-commitment contract. Furthermore, it has been shown that situations exist where the asset provider prefers to offer fixed-commitment and not capacity-option contracts. These situations include favorable expectations about the spot price and highly price-elastic demand in the contract market. The size of the Pareto improvement and thus the value of capacity-option contracts is driven especially by demand uncertainty in the contract market which capacity-option contracts can handle better than fixed-commitment contracts can.

The major source of the Pareto improvement is the ability of the split-tariff scheme to move a greater share of the buyer's capacity procurement to the contract market. With the reservation fee in the capacity-option contract being regularly below the capacity price in the comparable fixed-commitment contract, the buyer's cost trade-off is changed, inducing him to reserve a greater amount of capacity.

With regard to market efficiency, this feature is of even greater importance if the asset provider is able to realize cost savings through capacity provision on the basis of long-term contracts through earlier information on market demand derived from advance sale of capacity in the contract market. Earlier demand information may improve the seller's planning processes with regard to capacity allocation and network planning and might as well, because of a prolonged planning horizon, render more efficient hedging against fuel price and exchange rate risk possible. However, the results presented in Chap. 5 show that even without a cost advantage of the contract market, capacity options are in many cases beneficial for both buyer and seller.

Considering the potentially greater share of capacity reservations when capacity-option instead of conventional fixed-commitment contracts are being used, it has been shown in Chap. 6 how a seller of capacity options can further increase the expected profit of his capacity sales by anticipating non-execution of options and deliberately overbooking the non-reserved share of total capacity. To this end, an overbooking model for option-based advance sale of capacity has been developed.

By applying the capacity-option pricing model to a data set provided by Lufthansa Cargo AG, the applicability of capacity-option contracts for different routes of the carrier's flight schedule has been investigated in Chap. 7. The results of this application case study indicate that, depending on the characteristics of contract and spot market, capacity options are indeed on some flights performing better than fixed-commitment contracts, while on others the carrier should prefer to offer fixed-commitment contracts only. Furthermore, a first attempt to the monetary quantification of the financial merits of capacity-option contracts has yielded encouraging results.

8.1 Managerial Implications and Implementation Issues

The major obstacle with current contracting practices in the air cargo industry that has been identified in Chap. 2 is constituted by forwarders defaulting from contracts, be it because they simply face less demand from shippers than they have reserved capacity or be it because that capacity is available on the spot market at lower-than-contract rates.

By using capacity-option contracts, forwarders are explicitly given the choice to exercise some, all, or none of the purchased capacity options, rendering defaulting ex post from the contract pointless, because the reservation payment is due ex ante at signing the contract. Potentially, this can facilitate contract enforceability by the carrier. Managers negotiating a capacity-option contract should therefore be anxious to collect the reservation payment – or the present value thereof – actually prior to the start of the contract's life-span.

However, it remains to be seen if forwarders, considering current industry practices and difficulties with regard to contract enforceability, are willing to accept an upfront reservation payment. On the one hand, the preceding chapters provide evidence for the joint optimality of option contracts, i.e., regularly making buyer and seller better off, that should convince both parties to enter an agreement of this kind. On the other hand, capacity-option contracts clearly put a price-tag on something forwarders today oftentimes get for free, namely flexibility. The introduction and acceptance of option contracts might thus be facilitated if accompanied by realizing that flexibility indeed has a value and should as such be associated with a willingness to pay.

The thesis also provides insights concerning the selection of flights which are suitable for the introduction of capacity-option contracts. In the application case study carried out in Chap. 7, especially flights on which the discount for advance sale of capacity did not amount to more than approximately one third, appeared to be candidates for selling capacity options. The highest revenue impact was observed on flights that exhibited at the same time a comparably high demand from forwarders in the contract market.

Though the capacity-option pricing model stylizes the contract negotiation process between the buyer and seller to the most basic setting of the seller announcing the tariff and the buyer, given the seller's tariff, deciding on the amount of capacity units to reserve, the model provides also valuable insights for managers who actually negotiate capacity reservation contracts in a multi-stage negotiation process. It has been shown that the asset provider can attain the same, though suboptimal, profit level by multiple combinations of reservation and execution fee, providing managers both room to negotiate as well as a tool to steer the number of reservations received. This might be interesting if the seller's objective is to attain a certain target profit level and capacity utilization. In general, capacity-option contracts can be negotiated and closed in the same way as fixed-commitment contract, e.g., in a bilateral negotiation as common practice today. However, one might also think of using electronic market platforms, which potentially offer advantages by reducing transaction costs and increasing market liquidity.

Transaction costs for administrating long-term contractual agreements are not considered by the model. Insofar, one caveat with regard to the model results concerns the transaction costs associated with a split-tariff pricing scheme. If these should be higher than for a fixed-commitment contract (though it has been argued above that it is potentially easier to enforce a capacity-option than a fixed-commitment contract, which may lower the part of transaction costs related to contract enforcement), the financial improvement from using capacity options could be used up by increased transaction cost.

The model assumption of negligible transaction costs might however be justified if one considers the upcoming usage of electronic market platforms for settling and clearing long-term capacity agreements. The role of B2B platforms, however, goes well beyond facilitating transactions. If electronic marketplaces succeed in increasing the transparency and liquidity of contract and especially spot market, the quest for flexible capacity agreements that allow the buyer to react to favorable movements of the spot market might grow even stronger and thus accelerate the introduction of contingent contract types as, for example, capacity-option contracts.

While capacity-option contracts allow for effective risk sharing and can as such be regarded as a risk management tool, it has been shown in Chap. 5 that capacity options are not a panacea against uncertainty, but rather can help to deal with uncertainty better than fixed-commitment contracts. Furthermore, companies employing capacity-option contracts as a risk management tool should be aware that capacity-option contracts as a form of operational hedge should be seen as a complement rather than a substitute of financial hedging instruments (cf. Allayannis and Ihrig 2001), potentially including, for the example of an air cargo carrier, financial derivatives on fuel and foreign exchange.

8.2 Further Applications and Future Research

As a means of collaborative, but nevertheless competitive supply chain coordination, capacity options are part of the most recent literature on supply chain contracting. It has been discussed in Chap. 3 that there is a trend in supply chains for physical products towards flexible supply contracts. Though the focus in Chap. 3 has been on the more technical side of supply contracting – laying the theoretic foundations for the development of flexible capacity-option contracts –, the discussion also bears important implications for the air cargo supply chain: As established in Chap. 2, the demand for air cargo is a derived demand, depending on the underlying physical product flow actually constituting the cargo. In many instances, the shipper and consignee of an air cargo shipment are themselves partners in a physical-goods supply chain. If in this supply chain flexible contracts are used in order to accommodate fluctuations in demand for the raw material, component, or product at hand, it is clear that the transportation contracts within this physical-goods supply chain should exhibit a similar degree of flexibility in

| Supplier (Manufacturer) | Intermediary (Forwarder, 3PL) | Asset provider (Carrier) | Intermediary (Forwarder, 3PL) | Buyer (Retailer) |

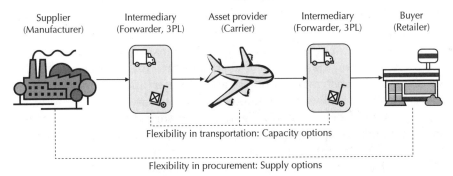

Flexibility in transportation: Capacity options

Flexibility in procurement: Supply options

Fig. 8.1. Capacity-option contracts in the context of a physical-goods supply chain employing flexible supply contracts

order to increase the efficiency of the entire system (see Fig. 8.1). Capacity-option contracts that allow for this flexibility thus represent the next logical step towards a holistic approach to supply chain management, aiming at system-wide performance improvement.

It has been shown in the preceding chapters how capacity options can be applied within the context of a revenue management system that serves price-sensitive demand from intermediaries in an advance-sale contract market and other groups of customers with potentially varying price-sensitivity in the spot market. While the model has been set up with capacity agreements between forwarding companies and airfreight carriers in mind, the model roles of intermediary and asset provider may also be cast differently.

Within the air cargo industry, a similar bulk-sale situation arises when two carriers, within the scope of an airline alliance, market capacity shares on each others' flights. While such capacity shares are today usually sold on the basis of transfer prices or exchanged mutually in the form of a barter transaction, the theoretical findings presented here suggest that capacity-option based pricing might be more suited to reflect the market value of a capacity share, given the heterogeneity of capacity values with respect to route, market, and time (e.g., day of the week) observed empirically (though admittedly other, e.g., strategic, reasons may be in the foreground that may call for non-market-based pricing of capacity shares).

For the evaluation of the model's applicability, the following characteristics are essential, supplementing the characteristics of industries apt for the successful application of revenue management practices in general (see Sec. 3.3.2): the presence of a contract market for advance bulk sale of capacity (or, more generally, the non-storable good or service at hand) to intermediaries acting as demand consolidators and/or resellers, considerable demand uncertainty faced by the intermediary, and price volatility in the spot market.

Beyond the air cargo, but still within the transportation industry, other modes of transport exhibit similar characteristics. The findings of this thesis should be transferable to other capacity reservation contracts, e.g., to container ship or railway operators who sell capacity in bulk to forwarding companies. Going beyond freight, the tourism industry might prove an interesting field of application. In a preliminary, empirical study, Hormann (2002) reports on contracting practices between hoteliers, (charter) airlines ("asset providers") and tour operators ("intermediaries") that today exhibit similar inefficiencies as observed in the air cargo industry. Since charter carriers and hotels seldom are in the position to enforce cancellation fees, tour operators can attain flexibility de facto for free and tend to overstate their capacity reservations, resulting often in renegotiations and short-term cancellation. Formulating the contractual arrangements between tour operators and asset providers as capacity options could potentially contribute to alleviating some of these problems.

In addition to investigating the application of capacity options and the capacity-option pricing model to the above mentioned industries, future research should also be directed at enlarging the empirical basis for the estimation of the financial impact of capacity options. The application case study reported on in Chap. 7 constitutes only a first step into this direction. However, it can be learned from this first attempt that more reliable and exact conclusions can probably only been drawn provided a more detailed and sound data basis. This would include a closer tracking of capacity drawn from contracts as well as of spot market sales and yield in order to obtain a set of historic data on which to base pricing decisions for a future offering of capacity-option contracts.

A

Mathematical Formulae

A.1 Normal Distribution

The following definitions and properties of the normal distribution in general and standardized form are used within the model presented in Chap. 4 and to derive the results presented in Chap. 5.

A.1.1 General Form

Let x be a random variable with expected value $\mathcal{E}[x] = \mu$ and variance $\mathrm{Var}[x] = \sigma^2$. Variable x is said to be normally distributed (denoted by $x \sim N(\mu, \sigma^2)$), if the distribution density function $f(x)$ is

$$f(x) = \frac{1}{\sigma\sqrt{2\pi}} e^{-\frac{(x-\mu)^2}{2\sigma^2}} \qquad (A.1)$$

The cumulative distribution function $F(x)$ is defined as:

$$F(x) = \int_{-\infty}^{x} f(y)\,dy \qquad (A.2)$$

(cf. Rinne 1997, p. 355). It can be expressed using the error function as

$$F(x) = \frac{1}{2}\left[1 + \mathrm{erf}\left(\frac{x-\mu}{\sigma\sqrt{2}}\right)\right] \qquad (A.3)$$

(cf. Abramowitz and Stegun 1972, p. 934).

A.1.2 Standard Normal Distribution

Define a random standardized variable z with

$$z = \frac{x - \mu}{\sigma} \tag{A.4}$$

and expected value $\mathcal{E}[z] = 0$ and variance $\text{Var}[z] = 1$.

The probability density function of z is the standard normal distribution density function $\varphi(z)$, defined as:

$$\varphi(z) = \frac{1}{\sqrt{2\pi}} e^{-\frac{z^2}{2}} \tag{A.5}$$

The standard normal cumulative distribution function $\Phi(z)$ is:

$$\Phi(z) = \int_{-\infty}^{z} f(x) dx \tag{A.6}$$

If follows for the relation between standard normal density and general normal density function that:

$$f(x) = \frac{1}{\sigma} \varphi\left(\frac{x - \mu}{\sigma}\right) \quad \text{and} \quad \varphi(z) = \sigma f(\mu + z\sigma) \tag{A.7}$$

Likewise, for the relation between standard normal cumulative and general normal cumulative distribution function, it holds true that:

$$F(x) = \Phi\left(\frac{x - \mu}{\sigma}\right) \quad \text{and} \quad \Phi(z) = F(\mu + z\sigma) \tag{A.8}$$

(cf. Rinne 1997, p. 355).

A.1.3 First Partial Moment

Let $z_a = (a - \mu)/\sigma$. For $x \sim N(\mu, \sigma^2)$, Winkler et al. (1972) show that the first partial moment can be determined by the following equation:

$$\int_{-\infty}^{a} x f(x) dx = \mu \Phi(z_a) - \sigma \varphi(z_a) \tag{A.9}$$

A.1.4 Standard Normal Loss Function

Based on (A.9), the standard normal loss function $L(z)$ is defined as:

$$L(z_a) = \int_{a}^{\infty} (x - a) \varphi(x) dx = \varphi(z_a) - [1 - \Phi(z_a)] z_a \tag{A.10}$$

(cf. Cachon and Terwiesch 2003, p. 258; Nahmias 2001, p. 254).

A.1.5 Partial Expectations

Let again $z_a = (a - \mu)/\sigma$ and $x \sim N(\mu, \sigma^2)$. Using the definition in (A.10), it follows from (A.9) in combination with (A.7) that the partial expectations $\mathcal{E}[(x - a)^+]$ and $\mathcal{E}[(a - x)^+]$ can be calculated in the following ways:

$$\mathcal{E}\left[(x - a)^+\right] = \int_a^\infty (x - a)f(x)\mathrm{d}x = \sigma L(z_a) \qquad (A.11)$$

$$\mathcal{E}\left[(a - x)^+\right] = \int_{-\infty}^a (a - x)f(x)\mathrm{d}x = \sigma\left[L(z_a) + z_a\right] \qquad (A.12)$$

Furthermore, it can be shown that

$$\mathcal{E}[\min(x, a)] = \int_{-\infty}^a xf(x)\mathrm{d}x + \int_a^\infty af(x)\mathrm{d}x = \mu - \sigma L(z_a) \qquad (A.13)$$

A.1.6 Primitive of the Cumulative Distribution Function

Let $x \sim N(\mu, \sigma^2)$ and $F(x)$ denote the cumulative normal distribution function. Using the relation in (A.3), the primitive of $F(x)$, denoted by $\bar{F}(x)$, is

$$\bar{F}(x) = \frac{e^{-\frac{(x-\mu)^2}{2\sigma^2}}\sigma}{\sqrt{2\pi}} + \frac{x}{2} + \frac{1}{2}(x - \mu)\mathrm{erf}\left(\frac{x - \mu}{\sigma\sqrt{2}}\right). \qquad (A.14)$$

A.2 Leibniz' Rule

Leibniz' rule is used for the differentiation of a definite integral whose limits are functions of the differential variable:

$$\frac{\partial}{\partial z}\int_{a(z)}^{b(z)} f(x, z)\mathrm{d}x = \int_{a(z)}^{b(z)} \frac{\partial f}{\partial z}\mathrm{d}x + f(b(z), z)\frac{\partial b}{\partial z} - f(a(z), z)\frac{\partial a}{\partial z} \qquad (A.15)$$

(cf. Meyberg and Vachenauer 1993, p.432).

B

Proofs and Calculations

B.1 Second Partial Derivatives of the Asset Provider's Expected Profit

B.1.1 With Respect to the Reservation Fee

From (4.30) with (4.31)–(4.33) it follows that the second partial derivative of Π_A with respect to r is given by:

$$\frac{\partial^2 \Pi_A}{\partial r^2} = \frac{\partial}{\partial r}\left[N + (r - c)\frac{\partial N}{\partial r} - (x - v)(1 - \Phi(z_x))\left(\frac{\sigma_{\tilde{D}_C}(1 - \Phi(z_N))}{\varphi(z_N)\sigma_{\tilde{s}}L(z_x)} + b \right) \right.$$

$$\left. - (\mathcal{E}[\tilde{s}] - t)(1 - \Phi(z_B))\frac{\partial N}{\partial r} \right] \tag{B.1}$$

with

$$\frac{\partial}{\partial r}\left[N + (r - c)\frac{\partial N}{\partial r} \right] = 2\frac{\partial N}{\partial r} + (r - c)\frac{\partial^2 N}{\partial r^2} \tag{B.2a}$$

$$\frac{\partial}{\partial r}\left[(x - v)(1 - \Phi(z_x))\left(\frac{\sigma_{\tilde{D}_C}(1 - \Phi(z_N))}{\varphi(z_N)\sigma_{\tilde{s}}L(z_x)} + b \right) \right] =$$

$$(x - v)(1 - \Phi(z_x))\frac{\sigma_{\tilde{D}_C}[z_N(1 - \Phi(z_N)) - \varphi(z_N)]}{[\varphi(z_N)\sigma_{\tilde{s}}L(z_x)]^2} \tag{B.2b}$$

$$\frac{\partial}{\partial r}\left[(\mathcal{E}[\tilde{s}] - t)(1 - \Phi(z_B))\frac{\partial N}{\partial r} \right] =$$

$$(\mathcal{E}[\tilde{s}] - t)\left[\frac{\varphi(z_B)}{\sigma_{\tilde{D}_S}}\left(\frac{\partial N}{\partial r} \right)^2 + (1 - \Phi(z_B))\frac{\partial^2 N}{\partial r^2} \right] \tag{B.2c}$$

and

$$\frac{\partial^2 N}{\partial r^2} = \frac{\sigma_{\bar{D}_C} z_N}{[\varphi(z_N)\sigma_{\bar{s}} L(z_x)]^2} \tag{B.3}$$

Substituting (B.2a)–(B.2c) and (B.3) into (B.1), with $\partial N/\partial r$ from (4.17a), and rearranging terms gives:

$$\frac{\partial^2 \Pi_A}{\partial r^2} = \left[z_N \sigma_{\bar{D}_S} \left[r - c - (x-v)(1-\Phi(z_x))(1-\Phi(z_N)) - (\mu_{\bar{s}} - t)(1-\Phi(z_B)) \right] \right.$$

$$\left. + (x-v)(1-\Phi(z_x))\varphi(z_N)\sigma_{\bar{D}_S} - (\mu_{\bar{s}} - t)\varphi(z_B)\sigma_{\bar{D}_C} \right] \frac{\sigma_{\bar{D}_C}}{\sigma_{\bar{D}_S}[\varphi(z_N)\sigma_{\bar{s}} L(z_x)]^2}$$

$$- \left[\sigma_{\bar{D}_S} + (\mu_{\bar{s}} - t)\varphi(z_B)b \right] \frac{2\sigma_{\bar{D}_C}}{\sigma_{\bar{D}_S}\varphi(z_N)\sigma_{\bar{s}} L(z_x)} - \left[2\sigma_{\bar{D}_S} + (\mu_{\bar{s}} - t)\varphi(z_B)b \right] \frac{b}{\sigma_{\bar{D}_S}} \tag{B.4}$$

B.1.2 With Respect to the Execution Fee

From (4.34) with (4.35)–(4.36) it follows that the second partial derivative of Π_A with respect to x is given by:

$$\frac{\partial^2 \Pi_A}{\partial x^2} = \frac{\partial}{\partial x} \left[(r-c)\frac{\partial N}{\partial x} + (\mu_{\bar{D}_C} - \sigma_{\bar{D}_C} L(z_N)) \left((1-\Phi(z_x)) - (x-v)\frac{\varphi(z_x)}{\sigma_{\bar{s}}} \right) \right.$$

$$- (x-v)(1-\Phi(z_x)) \left(b + \frac{\sigma_{\bar{D}_C}(1-\Phi(z_N))^2(1-\Phi(z_x))}{\varphi(z_N)\sigma_{\bar{s}} L(z_x)} \right)$$

$$\left. - (\mu_{\bar{s}} - t)(1-\Phi(z_B))\frac{\partial N}{\partial x} \right] \tag{B.5}$$

with

$$\frac{\partial}{\partial x}\left[(r-c)\frac{\partial N}{\partial x} \right] = (r-c)\frac{\partial^2 N}{\partial x^2} , \tag{B.6a}$$

$$\frac{\partial}{\partial x}\left[(\mu_{\bar{D}_C} - \sigma_{\bar{D}_C} L(z_N)) \left((1-\Phi(z_x)) - (x-v)\frac{\varphi(z_x)}{\sigma_{\bar{s}}} \right) \right] =$$

$$\left[b + \frac{\sigma_{\bar{D}_C}(1-\Phi(z_N))^2(1-\Phi(z_x))}{\varphi(z_N)\sigma_{\bar{s}} L(z_x)} \right] \left[(x-v)\frac{\varphi(z_x)}{\sigma_{\bar{s}}} - (1-\Phi(z_x)) \right]$$

$$+ \frac{\varphi(z_x)}{\sigma_{\bar{s}}} \left[\mu_{\bar{D}_C} - \sigma_{\bar{D}_C} L(z_N) \right] \left[(x-v)\frac{z_x}{\sigma_{\bar{s}}} - 2 \right] , \tag{B.6b}$$

$$\frac{\partial}{\partial x}\left[(x-v)(1-\Phi(z_x))\left(b+\frac{\sigma_{\tilde{D}_C}(1-\Phi(z_N))^2(1-\Phi(z_x))}{\varphi(z_N)\sigma_{\tilde{s}}L(z_x)}\right)\right]=$$

$$-b\left[\frac{\varphi(z_x)}{\sigma_{\tilde{s}}}-(1-\Phi(z_x))\right]+\left[\frac{\partial N}{\partial x}+b\right]\left[1-\Phi(z_N)\right]\left[(x-v)\frac{\varphi(z_x)}{\sigma_{\tilde{s}}}-(1-\Phi(z_x))\right]$$

$$-(x-v)(1-\Phi(z_x))\left[(1-\Phi(z_N))\frac{\partial^2 N}{\partial x^2}+\frac{(1-\Phi(z_N))(1-\Phi(z_x))}{\sigma_{\tilde{s}}L(z_x)}\left(\frac{\partial N}{\partial x}+b\right)\right],$$

$$\text{(B.6c)}$$

$$\frac{\partial}{\partial x}\left[(\mu_{\tilde{s}}-t)(1-\Phi(z_B))\frac{\partial N}{\partial x}\right]=$$

$$(\mu_{\tilde{s}}-t)\left[\frac{\varphi(z_B)}{\sigma_{\tilde{D}_S}}\left(\frac{\partial N}{\partial x}\right)^2+(1-\Phi(z_B))\frac{\partial^2 N}{\partial x^2}\right],\quad\text{(B.6d)}$$

and, applying the definition in (A.10),

$$\frac{\partial^2 N}{\partial x^2}=$$

$$-\frac{\sigma_{\tilde{D}_C}(1-\Phi(z_N))\left[(1-\Phi(z_x))^2[L(z_N)+\varphi(z_N)]-\varphi(z_N)\varphi(z_x)L(z_x)\right]}{[\varphi(z_N)\sigma_{\tilde{s}}L(z_x)]^2}.$$

$$\text{(B.7)}$$

Substituting (B.6a)–(B.6d) and (B.7) into (B.5), with $\partial N/\partial x$ from (4.18a), and rearranging terms gives:

$$\frac{\partial^2 \Pi_A}{\partial x^2}=\left[r-c-(\mu_{\tilde{s}}-t)(1-\Phi(z_B))+(x-v)(1-\Phi(z_N))(1-\Phi(z_x))\right]\frac{\partial^2 N}{\partial x^2}$$

$$+\left[b+\frac{2\sigma_{\tilde{D}_C}(1-\Phi(z_N))^2(1-\Phi(z_x))}{\varphi(z_N)\sigma_{\tilde{s}}L(z_x)}\right]\left[(x-v)\frac{\varphi(z_x)}{\sigma_{\tilde{s}}}-(1-\Phi(z_x))\right]$$

$$+b\left[\frac{\varphi(z_x)}{\sigma_{\tilde{s}}}-(1-\Phi(z_x))\right]\frac{\varphi(z_x)}{\sigma_{\tilde{s}}}\left[\mu_{\tilde{D}_C}-\sigma_{\tilde{D}_C}L(z_N)\right]\left[(x-v)\frac{z_x}{\sigma_{\tilde{s}}}-2\right]$$

$$-(x-v)\frac{\sigma_{\tilde{D}_C}(1-\Phi(z_N))^2(1-\Phi(z_x))^3}{\varphi(z_N)[\sigma_{\tilde{s}}L(z_x)]^2}-(\mu_{\tilde{s}}-t)\frac{\varphi(z_B)}{\sigma_{\tilde{D}_S}}\left(\frac{\partial N}{\partial x}\right)^2\quad\text{(B.8)}$$

B.2 Offload Quantity

B.2.1 Expected Value

In (6.2), the expected offload quantity is given by $\mathcal{E}[\tilde{Q}] = \mathcal{E}[(\tilde{M} + \tilde{E} - K)^+]$. With $\tilde{M} = \min(\tilde{D}_S, B)$, this results in

$$\mathcal{E}[\tilde{Q}] = \mathcal{E}[\max(\min(\tilde{D}_S, B) + \tilde{E} - K, 0)] \tag{B.9}$$

where $\tilde{E}$ is given by (4.6). The following cases can then be distinguished:

$$\tilde{Q} = \begin{cases} (B - K)^+ & \text{if } \tilde{D}_S > B \text{ and } \tilde{s} \leq x, \\ (B + N - K)^+ & \text{if } \tilde{D}_S > B, \tilde{s} > x, \text{ and } \tilde{D}_C > N, \\ (B + \tilde{D}_C - K)^+ & \text{if } \tilde{D}_S > B, \tilde{s} > x, \text{ and } \tilde{D}_C \leq N, \\ (\tilde{D}_S - K)^+ & \text{if } \tilde{D}_S \leq B \text{ and } \tilde{s} \leq x, \\ (\tilde{D}_S + N - K)^+ & \text{if } \tilde{D}_S \leq B, \tilde{s} > x, \text{ and } \tilde{D}_C > N, \\ (\tilde{D}_S + \tilde{D}_C - K)^+ & \text{if } \tilde{D}_S \leq B, \tilde{s} > x, \text{ and } \tilde{D}_C \leq N. \end{cases} \tag{B.10}$$

It follows for the expected value:

$$\mathcal{E}[\tilde{Q}] = \int\limits_{\tilde{D}_S=B}^{\infty} \int\limits_{\tilde{s}=0}^{x} (B - K)^+ f_S(\tilde{D}_S) g(\tilde{s}) \mathrm{d}\tilde{D}_S \mathrm{d}\tilde{s}$$

$$+ \int\limits_{\tilde{D}_S=B}^{\infty} \int\limits_{\tilde{s}=x}^{\infty} \int\limits_{\tilde{D}_C=N}^{\infty} (B + N - K)^+ f_S(\tilde{D}_S) g(\tilde{s}) f_C(\tilde{D}_C) \mathrm{d}\tilde{D}_S \mathrm{d}\tilde{s} \mathrm{d}\tilde{D}_C$$

$$+ \theta(B + N - K) \int\limits_{\tilde{D}_S=B}^{\infty} \int\limits_{\tilde{s}=x}^{\infty} \int\limits_{\tilde{D}_C=K-B}^{N} (\tilde{D}_C - (K - B)) f_S(\tilde{D}_S) g(\tilde{s}) f_C(\tilde{D}_C) \mathrm{d}\tilde{D}_S \mathrm{d}\tilde{s} \mathrm{d}\tilde{D}_C$$

$$+ \theta(B - K) \int\limits_{\tilde{D}_S=K}^{B} \int\limits_{\tilde{s}=0}^{x} (\tilde{D}_S - K) f_S(\tilde{D}_S) g(\tilde{s}) \mathrm{d}\tilde{D}_S \mathrm{d}\tilde{s}$$

$$+ \theta(B + N - K) \int\limits_{\tilde{D}_S=K-N}^{B} \int\limits_{\tilde{s}=x}^{\infty} \int\limits_{\tilde{D}_C=N}^{\infty} (\tilde{D}_S - (K - N)) f_S(\tilde{D}_S) g(\tilde{s}) f_C(\tilde{D}_C) \mathrm{d}\tilde{D}_S \mathrm{d}\tilde{s} \mathrm{d}\tilde{D}_C$$

$$+ \theta(B + N - K) \int\limits_{\tilde{D}_S=K-\tilde{D}_C}^{B} \int\limits_{\tilde{s}=x}^{\infty} \int\limits_{\tilde{D}_C=K-B}^{N} (\tilde{D}_S + \tilde{D}_C - K) f_S(\tilde{D}_S) g(\tilde{s}) f_C(\tilde{D}_C) \mathrm{d}\tilde{D}_S \mathrm{d}\tilde{s} \mathrm{d}\tilde{D}_C \tag{B.11}$$

where $\theta(x)$ is the Heaviside step function (also known as unit step function; Abramowitz and Stegun 1972, p. 1020) and defined by

$$\theta(x) = \frac{1}{2}\left[1 + \text{sgn}(x)\right] = \begin{cases} 0 & \text{if } x < 0, \\ \frac{1}{2} & \text{if } x = 0, \\ 1 & \text{if } x > 0. \end{cases} \tag{B.12}$$

The first five addends can be formulated analytically in near-closed form and calculated for the case of normally distributed variables assumed in the model:

$$\begin{aligned}
\mathcal{E}[\tilde{Q}] &= [1 - F_S(B)]G(x)(B - K)^+ \\
&+ [1 - F_S(B)][1 - G(x)][1 - F_C(N)](B + N - K)^+ \\
&+ \theta(B + N - K)[1 - F_S(B)][1 - G(x)][(B + N - K)F_C(N) - \bar{F}_C(N) + \bar{F}_C(K - B)] \\
&+ \theta(B - K)G(x)[(B - K)F_S(B) - \bar{F}_S(B) + \bar{F}_S(K)] \\
&+ \theta(B + N - K)[1 - G(x)][1 - F_C(N)][(B + N - K)F_S(B) - \bar{F}_S(B) + \bar{F}_S(K - N)] \\
&+ \theta(B + N - K)[1 - G(x)] \int_{\tilde{D}_S = K - \tilde{D}_C}^{B} \int_{\tilde{D}_C = K - B}^{N} (\tilde{D}_S + \tilde{D}_C - K)f_S(\tilde{D}_S)f_C(\tilde{D}_C)\mathrm{d}\tilde{D}_S\mathrm{d}\tilde{D}_C
\end{aligned} \tag{B.13}$$

where $\bar{F}(a)$ is the primitive of the cumulative distribution function and defined by

$$\bar{F}(a) = \int_{\infty}^{a} F(x)\mathrm{d}x. \tag{B.14}$$

The primitive of the cumulative normal distribution function is given in (A.14).

However, the sixth addend in (B.11) and (B.13) cannot be expressed in near-closed form.

B.2.2 Derivative with Respect to the Booking Level

The derivative of $\mathcal{E}[\tilde{Q}]$ with respect to B is determined by applying Leibniz' rule to (B.11).

Case 1 $K - N < B < K$

$$\begin{aligned}
\frac{\partial \mathcal{E}[\tilde{Q}]}{\partial B} &= \int_{\tilde{D}_S = B}^{\infty} \int_{\tilde{s} = x}^{\infty} \int_{\tilde{D}_C = N}^{\infty} f_S(\tilde{D}_S)g(\tilde{s})f_C(\tilde{D}_C)\mathrm{d}\tilde{D}_S\mathrm{d}\tilde{s}\mathrm{d}\tilde{D}_C \\
&- \int_{\tilde{s} = x}^{\infty} \int_{\tilde{D}_C = N}^{\infty} (B + N - K)f_S(B)g(\tilde{s})f_C(\tilde{D}_C)\mathrm{d}\tilde{s}\mathrm{d}\tilde{D}_C
\end{aligned}$$

$$+ \int_{\tilde{D}_S=B}^{\infty} \int_{\tilde{s}=x}^{\infty} \int_{\tilde{D}_C=K-B}^{N} f_S(\tilde{D}_S)g(\tilde{s})f_C(\tilde{D}_C)d\tilde{D}_S d\tilde{s} d\tilde{D}_C$$

$$- \int_{\tilde{s}=x}^{\infty} \int_{\tilde{D}_C=K-B}^{N} (\tilde{D}_C - K + B)f_S(B)g(\tilde{s})f_C(\tilde{D}_C)d\tilde{s}d\tilde{D}_C$$

$$+ \int_{\tilde{s}=x}^{\infty} \int_{\tilde{D}_C=N}^{\infty} (B - K + N)f_S(B)g(\tilde{s})f_C(\tilde{D}_C)d\tilde{s}d\tilde{D}_C$$

$$+ \int_{\tilde{s}=x}^{\infty} \int_{\tilde{D}_C=K-B}^{N} (B + \tilde{D}_C - K)f_S(B)g(\tilde{s})f_C(\tilde{D}_C)d\tilde{s}d\tilde{D}_C \quad \text{(B.15)}$$

Simplifying and integrating yields:

$$\frac{\partial \mathcal{E}[\tilde{Q}]}{\partial B} = [1 - F_S(B)][1 - G(x)][1 - F_C(K - B)] \quad \text{(B.16)}$$

Case 2 $B > K$

$$\frac{\partial \mathcal{E}[\tilde{Q}]}{\partial B} = \int_{\tilde{D}_S=B}^{\infty} \int_{\tilde{s}=0}^{x} f_S(\tilde{D}_S)g(\tilde{s})d\tilde{D}_S d\tilde{s}$$

$$- \int_{\tilde{s}=0}^{x} (B - K)f_S(B)g(\tilde{s})d\tilde{s}$$

$$+ \int_{\tilde{D}_S=B}^{\infty} \int_{\tilde{s}=x}^{\infty} \int_{\tilde{D}_C=N}^{\infty} f_S(\tilde{D}_S)g(\tilde{s})f_C(\tilde{D}_C)d\tilde{D}_S d\tilde{s} d\tilde{D}_C$$

$$- \int_{\tilde{s}=x}^{\infty} \int_{\tilde{D}_C=N}^{\infty} (B + N - K)f_S(B)g(\tilde{s})f_C(\tilde{D}_C)d\tilde{s}d\tilde{D}_C$$

$$+ \int_{\tilde{D}_S=B}^{\infty} \int_{\tilde{s}=x}^{\infty} \int_{\tilde{D}_C=K-B}^{N} f_S(\tilde{D}_S)g(\tilde{s})f_C(\tilde{D}_C)d\tilde{D}_S d\tilde{s} d\tilde{D}_C$$

$$- \int_{\tilde{s}=x}^{\infty} \int_{\tilde{D}_C=K-B}^{N} (\tilde{D}_C - K + B)f_S(B)g(\tilde{s})f_C(\tilde{D}_C)d\tilde{s}d\tilde{D}_C$$

$$+ \int\limits_{\tilde{s}=0}^{x} (B-K)f_S(B)g(\tilde{s})\mathrm{d}\tilde{s}$$

$$+ \int\limits_{\tilde{s}=x}^{\infty} \int\limits_{\tilde{D}_C=N}^{\infty} (B-K+N)f_S(B)g(\tilde{s})f_C(\tilde{D}_C)\mathrm{d}\tilde{s}\mathrm{d}\tilde{D}_C$$

$$+ \int\limits_{\tilde{s}=x}^{\infty} \int\limits_{\tilde{D}_C=K-B}^{N} (B+\tilde{D}_C-K)f_S(B)g(\tilde{s})f_C(\tilde{D}_C)\mathrm{d}\tilde{s}\mathrm{d}\tilde{D}_C \quad (\text{B.17})$$

After simplifying, integrating, and with $1 - F_C(K-B) = 0$ for the case considered at present, this results in:

$$\frac{\partial \mathcal{E}[\tilde{Q}]}{\partial B} = [1 - F_S(B)]G(x). \quad (\text{B.18})$$

$\mathcal{E}[\tilde{Q}]$ is not differentiable at $B = K$ and $B = K - N$. The left- and right-hand limits at $B = K$ are:

$$\lim_{B \to K^-} \frac{\partial \mathcal{E}[\tilde{Q}]}{\partial B} = [1 - F_S(K)][1 - G(x)](1 - 0) \quad (\text{B.19})$$

$$\lim_{B \to K^+} \frac{\partial \mathcal{E}[\tilde{Q}]}{\partial B} = [1 - F_S(K)]G(x) \quad (\text{B.20})$$

and thus

$$\lim_{B \to K^-} \frac{\partial \mathcal{E}[\tilde{Q}]}{\partial B} \neq \lim_{B \to K^+} \frac{\partial \mathcal{E}[\tilde{Q}]}{\partial B}. \quad (\text{B.21})$$

The left- and right-hand limits at $B = K - N$ are:

$$\lim_{B \to (K-N)^-} \frac{\partial \mathcal{E}[\tilde{Q}]}{\partial B} = [1 - F_S(K-N)][1 - G(x)][1 - F_C(N)] \quad (\text{B.22})$$

$$\lim_{B \to (K-N)^+} \frac{\partial \mathcal{E}[\tilde{Q}]}{\partial B} = 0 \quad (\text{B.23})$$

and thus

$$\lim_{B \to (K-N)^-} \frac{\partial \mathcal{E}[\tilde{Q}]}{\partial B} \neq \lim_{B \to (K-N)^+} \frac{\partial \mathcal{E}[\tilde{Q}]}{\partial B}. \quad (\text{B.24})$$

References

Abramowitz, M. and I. A. Stegun (eds.) (1972): *Handbook of Mathematical Functions*. 10th ed., U.S. Government Printing Office, Washington, D.C.

Allayannis, G. and J. Ihrig (2001): Exchange-Rate Hedging: Financial versus Operational Strategies. *American Economic Review* 91(2), pp. 391–395.

Althen, W., M. Graumann, and M. Niedermayer (2001): Alternative Wettbewerbsstrategien von Fluggesellschaften in der Luftfrachtbranche. *Schmalenbachs Zeitschrift für betriebswirtschaftliche Forschung* 53(6), pp. 420–441.

Amram, M. and N. Kulatilaka (1999): *Real Options – Managing Strategic Investment in an Uncertain World*. HBS Press, Boston (Mass.).

Anderson, C. K., M. Davison, and H. Rasmussen (2004): Revenue Management: A Real Options Approach. *Naval Research Logistics* 51(5), pp. 686–703.

Arrow, K. J. (2002): The Genesis of "Optimal Inventory Policy". *Operations Research* 50(1), pp. 1–2.

Arrow, K. J., T. Harris, and J. Marschak (1951): Optimal inventory policy. *Econometrica* 19(3), pp. 250–272.

Baker, T. K. and D. A. Collier (2003): The Benefits of Optimizing Prices to Manage Demand in Hotel Revenue Management Systems. *Production & Operations Management* 12(4), pp. 502–518.

Barde, D. J. and J. K. Mueller (1999): New for the Millennium: 4PL. *Transportation & Distribution* 40(2), pp. 78–80.

Barnes-Schuster, D., Y. Bassok, and R. Anupindi (2002): Coordination and Flexibility in Supply Contracts with Options. *Manufacturing & Service Operations Management* 4(3), pp. 171–207.

Bazerman, M. H. and J. J. Gillespie (1999): Betting on the Future: The Virtues of Contingent Contracts. *Harvard Business Review* 77(5), pp. 155–160.

Beckmann, M. J. (1958): Decision and Team Problems in Airline Reservations. *Econometrica* 26(1), pp. 134–145.

Belobaba, P. P. (1987): Airline Yield Management: an Overview of Seat Inventory Control. *Transportation Science* 21(2), pp. 63–73.

Belobaba, P. P. (1989): Application of a Probabilistic Decision Model to Airline Seat Inventory Control. *Operations Research* 37(2), pp. 183–197.

Belobaba, P. P. (1998): The Evolution of Airline Yield Management: Fare Class to Origin-Destination Seat Inventory Control. In: G. F. Butler and M. R. Keller (eds.), *Handbook of Airline Marketing*, chap. 23, McGraw-Hill, New York (N.Y.), pp. 285–302.

Berge, M. E. and C. A. Hopperstad (1993): Demand Driven Dispatch: a Method for Dynamic Aircraft Capacity Assignment, Models and Algorithms. *Operations Research* 41(1), pp. 153–168.

Bertsekas, D. P. (1999): *Nonlinear Programming*. 2nd ed., Athena Scientific, Belmont (Mass.).

Bertsimas, D. and I. Popescu (2003): Revenue Management in a Dynamic Network Environment. *Transportation Science* 37(3), pp. 257–277.

Bitran, G. and R. Caldentey (2003): An Overview of Pricing Models for Revenue Management. *Manufacturing & Service Operations Management* 5(3), pp. 203–229.

Black, F. and M. Scholes (1973): The Pricing of Options and Corporate Liabilities. *Journal of Political Economy* 81(3), pp. 637–654.

Blonski, M. (2002): Network externalities and two-part tariffs in telecommunication markets. *Information Economics and Policy* 14(1), pp. 95–109.

Bodily, S. E. and P. E. Pfeifer (1992): Overbooking Decision Rules. *Omega* 20(1), pp. 129–133.

Boeing (2002): World Air Cargo Forecast 2002/2003. Boeing Commercial Airplanes, Seattle (Wash.).

Boeing (2004): World Air Cargo Forecast 2004/2005. Boeing Commercial Airplanes, Seattle (Wash.).

de Boer, S. V., R. Freling, and N. Piersma (2002): Mathematical programming for network revenue management revisited. *European Journal of Operational Research* 137(1), pp. 72–92.

Botimer, T. C. and P. P. Belobaba (1999): Airline pricing and fare product differentiation: A new theoretical framework. *Journal of the Operational Research Society* 50(11), pp. 1085–1097.

Brealey, R. A. and S. C. Myers (2000): *Principles of Corporate Finance*. 6th ed., McGraw-Hill, Boston (Mass.).

Brumelle, S. L. and J. I. McGill (1993): Airline seat allocation with multiple nested fare classes. *Operations Research* 41(1), pp. 127–137.

Brzoska, L. (2003): *Die Conjoint-Analyse als Instrument zur Prognose von Preisreaktionen: eine theoretische und empirische Beurteilung der externen Validität*. Kovac, Hamburg.

Burnetas, A. and S. Gilbert (2001): Future Capacity Procurements Under Unknown Demand and Increasing Costs. *Management Science* 47(7), pp. 979–992.

Cachon, G. P. (2003): Supply Chain Coordination with Contracts. In: A. G. de Kok and S. C. Graves (eds.), *Supply Chain Management: Design, Coordination and Operation*, vol. 11 of *Handbooks in Operations Research and Management Science*, chap. 6, Elsevier, Amsterdam, pp. 229–339.

Cachon, G. P. and M. A. Lariviere (2002): *Supply Chain Coordination with Revenue-Sharing Contracts: Strengths and Limitations*. Working paper, The Wharton School and Kellogg School of Management, Philadelphia (Pa.) and Evanston (Ill.).

Cachon, G. P. and C. Terwiesch (2003): *Matching Supply with Demand: An Introduction to Operations Management*. McGraw-Hill, New York (N.Y.).

Carroll, W. J. and R. C. Grimes (1995): Evolutionary Change in Product Management: Experiences in the Car Rental Industry. *Interfaces* 25(5), pp. 84–104.

Chatwin, R. E. (1996): Multi-Period Airline Overbooking with Multiple Fare Classes. *Naval Research Logistics* 43(5), pp. 603–612.

Chatwin, R. E. (1998): Multiperiod Airline Overbooking with a Single Fare Class. *Operations Research* 46(6), pp. 805–819.

Chatwin, R. E. (1999): Continuous-Time Airline Overbooking with Time-Dependent Fares and Refunds. *Transportation Science* 33(2), pp. 182–191.

Chen, F. (2001): Market Segmentation, Advanced Demand Information, and Supply Chain Performance. *Manufacturing & Service Operations Management* 3(1), pp. 53–67.

Cheng, F., M. Ettl, G. Lin, M. Schwarz, and D. D. Yao (2002): *Flexible Supply Contracts via Options*. Working paper, revised november 2003, IBM T. J. Watson Research Center, Yorktown Heights (N.Y.).

Ciancimino, A., G. Inzerillo, S. Lucidi, and L. Palagi (1999): A Mathematical Programming Approach for the Solution of the Railway Yield Management Problem. *Transportation Science* 33(2), pp. 168–181.

Coase, R. H. (1937): The Nature of the Firm. *Economica* 4(November), pp. 386–405.

Cohen, M. A. and N. Agrawal (1999): An analytical comparison of long and short term contracts. *IIE Transactions* 31(8), pp. 783–796.

Cooper, W. L. (2002): Asymptotic Behavior of an Allocation Policy for Revenue Management. *Operations Research* 50(4), pp. 720–727.

Copeland, T. and V. Antikarov (2001): *Real Options – A Practitioner's guide*. Texere, New York (N.Y.).

Cox, J. C., S. A. Ross, and M. Rubinstein (1979): Option Pricing: A Simplified Approach. *Journal of Financial Economics* 7(3), pp. 229–263.

Coyne, L. (2003): Why We Need Air Cargo to Prosper. In: E. Boulton (ed.), *Business Briefing – Aviation Strategies: Challenges & Opportunities of Liberalization*, World Markets Series, International Civil Aviation Organization, World Market Research Centre, London.

Cross, R. G. (1997): *Revenue Management: Hard-core Tactics for Market Domination*. Broadway Books, New York (N.Y.).

Cross, R. G. (1998): Trends in Airline Revenue Management. In: G. F. Butler and M. R. Keller (eds.), *Handbook of Airline Marketing*, chap. 24, McGraw-Hill, New York (N.Y.), pp. 303–318.

DeLain, L. and E. O'Meara (2004): Building a business case for revenue management. *Journal of Revenue and Pricing Management* 2(4), pp. 368–377.

van Delft, C. and J.-P. Vial (2004): A practical implementation of stochastic programming: an application to the evaluation of option contracts in supply chains. *Automatica* 40(5), pp. 743–756.

Desiraju, R. and S. M. Shugan (1999): Strategic Service Pricing and Yield Management. *Journal of Marketing* 63(1), pp. 44–56.

Dixit, A. K. and R. S. Pindyck (1994): *Investment under Uncertainty*. Princeton University Press, Princeton.

Doganis, R. (2002): *Flying Off Course – The Economics of International Airlines*. 3rd ed., Routledge, London.

Edgeworth, F. Y. (1888): The Mathematical Theory of Banking. *Journal of the Royal Statistical Society* 51(1), pp. 113–127.

Emmons, H. and S. M. Gilbert (1998): Note. The Role of Returns Policies in Pricing and Inventory Decisions for Catalogue Goods. *Management Science* 44(2), pp. 276–283.

Eppen, G. D. and A. V. Iyer (1997): Backup Agreements in Fashion Buying—The Value of Upstream Flexibility. *Management Science* 43(11), pp. 1469–1484.

Eye for Transport (2003): Lufthansa Cargo adjusts rates to demand. Online publication, accessed on August 6, 2003, URL http://www.eyefortransport.com/index.asp?news=37963&nli=freight&ch=.

FAZ (2001): Lufthansa Cargo will Spediteure binden. *Frankfurter Allgemeine Zeitung*, issue of July 4, 2001, p. 22.

Feng, Y. and G. Gallego (2000): Perishable Asset Revenue Management with Markovian Time Dependent Demand Intensities. *Management Science* 46(7), pp. 941–956.

Feng, Y. and B. Xiao (1999): Maximizing Revenues of Perishable Assets With a Risk Factor. *Operations Research* 47(2), pp. 337–341.

Feng, Y. and B. Xiao (2000a): A Continuous-Time Yield Management Model with Multiple Prices and Reversible Price Changes. *Management Science* 46(5), pp. 644–657.

Feng, Y. and B. Xiao (2000b): Optimal Policies of Yield Management with Multiple Predetermined Prices. *Operations Research* 48(2), pp. 332–343.

Feng, Y. and B. Xiao (2000c): Revenue Management with Two Market Segments and Reserved Capacity for Priority Customers. *Advances in Applied Probability* 32(3), pp. 800–823.

Feng, Y. and B. Xiao (2001): A Dynamic Airline Seat Inventory Control Model and its Optimal Policy. *Operations Resarch* 49(6), pp. 938–949.

Financial Times (2004a): Lufthansa Cargo sets sights on job cuts for turnround. *Financial Times*, issue of August 26, 2004, p. 18.

Financial Times (2004b): Shipping industry slow to ride wave of global demand. *Financial Times*, issue of March 8, 2004, p. 14.

Gallego, G., S. Krishnamoorthy, and R. Phillips (2003): *Competitive Revenue Management with Forward and Spot Markets*. Working paper, Columbia University, New York (N.Y.).

Geraghty, M. K. and E. Johnson (1997): Revenue management saves National Car Rental. *Interfaces* 27(1), pp. 107–127.

GF-X (2003): Lufthansa Cargo expands service for GF-X customers. Online press release, accessed on May 7, 2003, URL http://www.gf-x.com/news/news_archiveApr03.html.

GF-X (2004): Global Freight Exchange: The neutral reservations system for the airfreight industry. Corporate brochure, accessed on October 6, 2004, URL http://www.gf-x.com/downloads/GF-X_brochure.pdf.

Green, P. E., A. M. Krieger, and Y. J. Wind (2001): Thirty Years of Conjoint Analysis: Reflections and Prospects. *Interfaces* 31(3 [2/2]), pp. 56–73.

Grin, B. (1998): Developments in Air Cargo. In: G. F. Butler and M. R. Keller (eds.), *Handbook of Airline Marketing*, chap. 7, McGraw-Hill, New York (N.Y.), pp. 75–93.

Gundepudi, P., N. Rudi, and A. Seidmann (2001): Forward Versus Spot Buying of Information Goods. *Journal of Management Information Systems* 18(2), pp. 107–131.

Harris, F. H. d. and J. P. Pinder (1995): A Revenue Management Approach to Demand Management and Order Booking in Assemble-to-Order Manufacturing. *Journal of Operations Management* 13(4), pp. 299–309.

Hellermann, R. and A. Huchzermeier (2002a): *Lufthansa Cargo AG: Capacity Reservation and Dynamic Pricing*. Case study no. 602-029-1 at European Case Clearing House, Otto Beisheim Graduate School of Management (WHU), Vallendar.

Hellermann, R. and A. Huchzermeier (2002b): *Lufthansa Cargo AG: Capacity Reservation and Dynamic Pricing*. Teaching note no. 602-029-8 at European Case Clearing House, Otto Beisheim Graduate School of Management (WHU), Vallendar.

Herrmann, N., M. Müller, and A. Crux (1998a): Pricing and Revenue Management Can Reshape Your Competitive Position in Today's Air Cargo Business. In: G. F. Butler and M. R. Keller (eds.), *Handbook of Airline Marketing*, chap. 30, McGraw-Hill, New York (N.Y.), pp. 387–410.

Herrmann, N., D. Trefzger, and A. Crux (1998b): Challenges for Tomorrow's Successful Air Freight Providers: Nothing Is as Permanent as Change. In: G. F. Butler and M. R. Keller (eds.), *Handbook of Airline Marketing*, chap. 12, McGraw-Hill, New York (N.Y.), pp. 141–154.

Ho, T.-H. and J. Zhang (2004): *Experimental Test of Solutions to the Double-Marginalization Problem – A Reference-Dependent Approach*. Working paper, University of California at Berkeley, Berkeley (Ca.).

Holloway, S. (2003): *Straight and level: practical airline economics*. 2nd ed., Ashgate Publishing, Hampshire.

Hormann, M. (2002): *Potential Field of Application of Capacity Options in the Tourism Industry*. Praxisarbeit, WHU, Vallendar.

Huchzermeier, A. and C. H. Loch (2001): Project Management Under Risk: Using the Real Options Approach to Evaluate Flexibility in R&D. *Management Science* 47(1), pp. 85–101.

Hull, J. C. (2003): *Options, Futures and Other Derivatives*. 5th ed., Prentice Hall, Upper Saddle River (N.J.).

Humair, S. (2001): *Yield Management for Telecommunication Networks: Defining a New Landscape*. Ph.D. dissertation, MIT, Sloan School of Management, Cambridge (Mass.).

Hyman, D. N. (1993): *Modern Microeconomics: Analysis and Applications*. 3rd ed., Irwin, Homewood (Ill.).

Ishii, H. and T. Konno (1998): A Stochastic Inventory Problem with Fuzzy Shortage Cost. *European Journal of Operational Research* 107(1), pp. 90–94.

Jaillet, P., E. I. Ronn, and S. Tompaidis (2004): Valuation of Commodity-Based Swing Options. *Management Science* 50(7), pp. 909–921.

Jensen, M. C. and W. H. Meckling (1976): Theory of the firm: Managerial behavior, agency costs and ownership structure. *Journal of Financial Economics* 3(4), pp. 305–360.

Jeuland, A. P. and S. M. Shugan (1983): Managing Channel Profits. *Marketing Science* 2(3), pp. 239–272.

Jeuland, A. P. and S. M. Shugan (1988): Reply to: Managing Channel Profits: Comment. *Marketing Science* 7(1), pp. 103–106.

Jones, P. (1999): Yield management in UK hotels: a systems analysis. *Journal of the Operational Research Society* 50(11), pp. 1111–1119.

Kadar, M. and J. Larew (2004): Securing the Future of Air Cargo. *Mercer on Travel and Transport* 10(1), pp. 3–9.

Karaesmen, I. and G. J. van Ryzin (2004): Overbooking with Subtitutable Inventory Classes. *Operations Research* 52(1), pp. 83–104.

Kasilingam, R. G. (1996): Air cargo revenue management: Characteristics and complexities. *European Journal of Operational Research* 96(1), pp. 36–44.

Kasilingam, R. G. (1997): An Economic Model for Air Cargo Overbooking Under Stochastic Capacity. *Computers & Industrial Engineering* 32(1), pp. 221–226.

Kay, D. (2003): It's Time to Set Air Cargo Free. Online publication, accessed on August 3, 2004, URL http://www.tiaca.org/articles/2004/03/19/DB1DC4B404684041B6BC6E9CBA1B86E6.asp.

Kester, W. C. (1984): Today options for tomorrow's growth. *Harvard Business Review* 62(2), pp. 153–160.

Khouja, M. (1996): The Newsboy Problem with Multiple Discounts Offered by Suppliers and Retailers. *Decision Sciences* 27(3), pp. 589–599.

Khouja, M. (1999): The Single-Period (News-Vendor) Problem: Literature Review and Suggestions for Future Research. *Omega* 27(5), pp. 537–553.

Kimes, S. E. (1989): Yield Management: A Tool for Capacity-Constrained Service Firms. *Journal of Operations Management* 8(4), pp. 348–363.

Kimes, S. E. (1994): Perceived Fairness of Yield Management. *Cornell Hotel and Restaurant Administration Quarterly* 35(1), pp. 22–29.

Kimes, S. E. (2000): Revenue Management on the Links: Applying Yield Management to the Golf-course Industry. *Cornell Hotel and Restaurant Administration Quarterly* 41(1), pp. 120–127.

Kimes, S. E. (2002): Perceived Fairness of Yield Management: An Update. *Cornell Hotel and Restaurant Administration Quarterly* 43(1), pp. 28–29.

Kimes, S. E. (2003): Revenue Management: A Retrospective. *Cornell Hotel and Restaurant Administration Quarterly* 44(5/6), pp. 131–138.

Kimes, S. E. and R. B. Chase (1998): The Strategic Levers of Yield Management. *Journal of Service Research* 1(2), pp. 156–166.

Kimes, S. E., R. B. Chase, S. Choi, P. Y. Lee, and E. N. Ngonzi (1998): Restaurant Revenue Management: Applying Yield Management to the Restaurant Industry. *Cornell Hotel and Restaurant Administration Quarterly* 39(3), pp. 32–39.

Kimes, S. E. and J. Wirtz (2003): Has Revenue Management Become Acceptable? *Journal of Service Research* 6(2), pp. 125–135.

Kleindorfer, P. R. and D. J. Wu (2003): Integrating Long- and Short-Term Contracting via Business-to-Business Exchanges for Capital-Intensive Technologies. *Management Science* 49(11), pp. 1597–1615.

Knight, F. H. (1921): *Risk, Uncertainty, and Profit*. Hart, Schaffner & Marx, Boston (Mass.).

Ladany, S. P. and A. Arbel (1991): Optimal cruise-liner passenger cabin pricing policy. *European Journal of Operational Research* 55(2), pp. 136–147.

Lahoti, A. (2002a): Cargo Revenue Management – Strategy and Tactics to win the 'new business war'. Online publication, accessed on March 13, 2002, URL http://www.eyefortransport.com/index.asp?news=26667.

Lahoti, A. (2002b): Why CEOs Should Care About Revenue Management. *OR/MS Today* 29(1), pp. 34–37.

Lariviere, M. A. (1999): Supply Chain Contracting and Coordination with Stochastic Demand. In: S. Tayur, M. Magazine, and R. Ganeshan (eds.), *Quantitative Models for Supply Chain Management*, Kluwer, Boston (Mass.), pp. 233–268.

Lariviere, M. A. and E. L. Porteus (2001): Selling to the Newsvendor: An Analysis of Price-Only Contracts. *Manufacturing & Service Operations Management* 3(4), pp. 293–305.

Lau, A. H.-L. and H.-S. Lau (1988): The Newsboy Problem With Price-Dependent Demand Distribution. *IIE Transactions* 20(2), pp. 168–175.

Lau, H.-S. (1997): Simple Formulas for the Expected Costs in the Newsboy Problem: An Educational Note. *European Journal of Operational Research* 100(3), pp. 557–561.

Lau, H.-S. and A. H.-L. Lau (1999): Manufacturer's pricing strategy and return policy for a single-period commodity. *European Journal of Operational Research* 116(2), pp. 291–304.

Lee, H. and S. Whang (2002): The Impact of the Secondary Market on the Supply Chain. *Management Science* 48(6), pp. 719–731.

Lee, H. L. and S. Nahmias (1993): Single-Product, Single-Location Models. In: S. C. Graves, A. H. G. R. Kan, and P. H. Zipkin (eds.), *Logistics of Production and Inventory*, vol. 4 of *Handbooks in Operations Research and Management Science*, chap. 1, North-Holland, Amsterdam, pp. 3–55.

Lee, K. S. and I. C. L. Ng (2001): Advanced sale of service capacities: a theoretical analysis of the impact of price sensitivity on pricing and capacity allocations. *Journal of Business Research* 54(3), pp. 219–225.

Lieb, R. C., R. A. Millen, and L. N. van Wassenhove (1993): Third-party logistics: A comparison of experienced American and European manufactures. *International Journal of Physical Distribution & Logistics Management* 23(6), pp. 35–44.

Luce, R. D. and J. W. Tukey (1964): Simultaneous conjoint measurement: A new type of fundamental measurement. *Journal of Mathematical Psychology* 1(1), pp. 1–27.

Luenberger, D. G. (1998): *Investment Science*. Oxford University Press, New York (N.Y.).

Lufthansa (2004): Annual Report 2003 of Lufthansa AG. Deutsche Lufthansa AG, Frankfurt (Main).

Lufthansa Cargo (2004): Annual Report 2003 of Lufthansa Cargo AG. Lufthansa Cargo AG, Frankfurt (Main).

Maglaras, C. and J. Meissner (2004): *Dynamic Pricing Strategies for Multi-Product Revenue Management Problems*. Working paper, Columbia University, New York (N.Y.).

Martinez-de-Albéniz, V. and D. Simchi-Levi (2002): *A Portfolio Approach to Procurement Contracts*. Working paper, Massachusetts Institute of Technology, Cambridge (Mass.).

Martinez-de-Albéniz, V. and D. Simchi-Levi (2003): *Competition in the Supply Option Market*. Working paper, Massachusetts Institute of Technology, Cambridge (Mass.).

McGill, J. I. and G. J. van Ryzin (1999): Revenue Management: Research Overview and Prospects. *Transportation Science* 33(2), pp. 233–256.

Mendelson, H. and T. I. Tunca (2004): Strategic Trading, Liquidity, and Information Acquisition. *Review of Financial Studies* 17(2), pp. 295–337.

Merton, R. C. (1973): Theory of rational option pricing. *Bell Journal of Economics & Management Science* 4(1), pp. 141–183.

Metters, R. and V. Vargas (1999): Yield Management for the Nonprofit Sector. *Journal of Service Research* 1(3), pp. 215–226.

Meyberg, K. and P. Vachenauer (1993): *Höhere Mathematik 1*. 2nd ed., Springer, Heidelberg.

Mills, E. S. (1959): Uncertainty and Price Theory. *Quarterly Journal of Economics* 73(1), pp. 116–130.

Moorthy, K. S. (1987): Managing Channel Profits: Comment. *Marketing Science* 6(4), pp. 375–379.

Murphy, P. R. and R. F. Poist (2000): Third-party logistics: Some user versus provider perspectives. *Journal of Business Logistics* 21(2), pp. 121–133.

Nahmias, S. (2001): *Production and Operations Analysis*. 4th ed., McGraw-Hill/Irwin, New York (N.Y.).

Nahmias, S. (2004): *Production and Operations Analysis*. 5th ed., McGraw-Hill, New York (N.Y.).

Nair, S. K. and R. Bapna (2001): An application of yield management for Internet Service Providers. *Naval Research Logistics* 48(5), pp. 348–362.

Oi, W. Y. (1971): A Disneyland Dilemma: Two-Part Tariffs for a Mickey Mouse Monopoly. *Quarterly Journal of Economics* 85(1), pp. 77–96.

Orkin, E. B. (1998): Wishful Thinking and Rocket Science: The Essential Matter of Calculating Unconstrained Demand for Revenue Management. *Cornell Hotel and Restaurant Administration Quarterly* 39(4), pp. 15–19.

Ott, J. (2003): Air Transport Cargo Redux – Airlines do better moving freight than passengers. *Aviation Week & Space Technology* 158(13), pp. 46–49.

Oum, T. H., W. G. Waters II, and J.-S. Yong (1990): *A Survey of Recent Estimates of Price Elasticities of Demand for Transport*. Working paper, available at URL http://www.worldbank.org/transport/publicat/inu-70.pdf, The World Bank, Infrastructure and Urban Development Department, Washington (D.C.).

Oum, T. H., W. G. Waters II, and J.-S. Yong (1992): Concepts of Price Elasticities of Transport Demand and Recent Empirical Estimates: An Interpretative Survey. *Journal of Transport Economics and Policy* 26(2), pp. 139–154.

Padmanabhan, V. and I. P. L. Png (1995): Return Policies: Make Money by Making Good. *Sloan Management Review* 37(1), pp. 65–72.

Pasternack, B. A. (1985): Optimal Pricing and Return Policies for Perishable Commodities. *Marketing Science* 4(2), pp. 166–176.

Petruzzi, N. C. and M. Dada (1999): Pricing and the Newsvendor Problem: A Review with Extensions. *Operations Research* 47(2), pp. 183–194.

Petruzzi, N. C. and M. Dada (2002): Dynamic Pricing and Inventory control with Learning. *Naval Research Logistics* 49(3), pp. 303–325.

Pfeifer, P. E. (1989): The Airline Discount Fare Allocation Problem. *Decision Sciences* 20(1), pp. 149–157.

Pigou, A. C. (1932): *The Economics of Welfare*. 4th ed., MacMillan, London.

Png, I. P. L. (1989): Reservations: Customer Insurance in the Marketing of Capacity. *Marketing Science* 8(3), pp. 248–264.

Pompeo, L. and T. Sapountzis (2002): Freight expectations. *The McKinsey Quarterly* 2002(2), pp. 90–99.

Porteus, E. L. (1990): Stochastic Inventory Models. In: D. P. Heyman and M. J. Sobel (eds.), *Stochastic Models*, vol. 2 of *Handbooks in Operations Research and Management Science*, chap. 12, North-Holland, Amsterdam, pp. 605–652.

Quan, D. C. (2002): The Price of a Reservation. *Cornell Hotel and Restaurant Administration Quarterly* 43(3), pp. 77–86.

Reece, W. S. and R. S. Sobel (2000): Diagrammatic Approach to Capacity-Constrained Price Discrimination. *Southern Economic Journal* 66(4), pp. 1001–1008.

Rinne, H. (1997): *Taschenbuch der Statistik*. 2nd ed., Verlag Harri Deutsch, Thun and Franfurt am Main.

Rudi, N. and D. F. Pyke (2000): Teaching Supply Chain Concepts with the Newsboy Model. In: M. E. Johnson and D. F. Pyke (eds.), *Supply Chain Management: Innovations for Education*, POMS, Miami, pp. 170–180.

Schmelter, L. (2002): *Produkt- und Preismanagement für Serviceangebote in der Luftfrachtindustrie: Integration von nachfrage- und wettbewerbsseitigen Marktforschungsanalysen am Beispiel der Lufthansa Cargo AG*. Diploma thesis, Leipzig Graduate School of Management (HHL), Leipzig.

Schneider, D. (1993): *Wettbewerbsvorteile integrierter Systemanbieter im Luftfrachtmarkt*. No. 1458 in Europäische Hochschulschriften, Peter Lang, Frankfurt.

Serel, D. A., M. Dada, and H. Moskowitz (2001): Sourcing decisions with capacity reservation contracts. *European Journal of Operational Research* 131(3), pp. 635–648.

Shields, M. (1998): The Changing Cargo Business. In: G. F. Butler and M. R. Keller (eds.), *Handbook of Airline Marketing*, chap. 15, McGraw-Hill, New York (N.Y.), pp. 183–187.

Shugan, S. M. and J. Xie (2000): Advance Pricing of Services and Other Implications of Separating Purchase and Consumption. *Journal of Service Research* 2(3), pp. 227–239.

Silver, E. A. and R. Peterson (1985): *Decision Systems for Inventory Management and Production Planning*. 2nd ed., John Wiley & Sons, New York.

Silver, E. A., D. F. Pyke, and R. Peterson (1998): *Inventory Management and Production Planning and Scheduling*. 3rd ed., John Wiley & Sons, New York (N.Y.).

Simon, H. (1992): *Preismanagement*. 2nd ed., Gabler, Wiesbaden.

Skiera, B. (1999): *Mengenbezogene Preisdifferenzierung bei Dienstleistungen*. Gabler, Wiesbaden.

Skjoett-Larsen, T. (2000): Third-party logistics – from an interorganizational point of view. *International Journal of Physical Distribution & Logistics Management* 30(2), pp. 112–121.

Slager, B. and L. Kapteijns (2004): Implementation of cargo revenue management at KLM. *Jounral of Revenue and Pricing Management* 3(1), pp. 80–90.

Smith, B. C., D. P. Günther, B. V. Rao, and R. M. Ratliff (2001): E-Commerce and Operations Research in Airline Planning, Marketing, and Distribution. *Interfaces* 31(2), pp. 37–55.

Smith, B. C., J. F. Leimkuhler, and R. M. Darrow (1992): Yield Management at American Airlines. *Interfaces* 22(1), pp. 8–31.

Sowinski, L. L. (2002): Air Cargo Encounters Major Turbulences in 2001. *World Trade* 15(5), pp. 30–32.

Spengler, J. J. (1950): Vertical Integration and Antitrust Policy. *The Journal of Political Economy* 58, pp. 347–352.

Spinler, S. (2003): *Capacity Reservation for Capital-Intensive Technologies: An Options Approach*. Springer, Heidelberg.

Spinler, S., A. Huchzermeier, and P. Kleindorfer (2003): Risk hedging via options contracts for physical delivery. *OR Spectrum* 25(3), pp. 379–395.

Spremann, K. and M. Klinkhammer (1985): Grundgebühren und zweiteilige Tarife. *Zeitschrift für Betriebswirtschaft* 55(8), pp. 790–820.

Spulber, D. F. (1992): Optimal Nonlinear Pricing and Contingent Contracts. *International Economic Review* 33(4), pp. 747–772.

Stonier, J. E. (1999): What Is an Aircraft Purchase Option Worth? Quantifying Asset Flexibility Created through Manufacturer Lead-Time Reductions and Product Commonality. In: G. F. Butler and M. R. Keller (eds.), *Handbook of Airline Finance*, McGraw-Hill, New York (N.Y.), pp. 231–250.

Stuhlmann, S. (2000): *Kapazitätsgestaltung in Dienstleistungsunternehmen: Eine Analyse aus Sicht des externen Faktors*. Gabler, Wiesbaden.

Swan, W. M. (2002): Airline demand distributions: passenger revenue management and spill. *Transportation Research Part E* 38(3–4), pp. 253–263.

Talley, W. K. and A. Schwarz-Miller (1988): The Demand for Air Services Provided by Air Passenger-Cargo Carriers in a Deregulated Environment. *International Journal of Transport Economics* 15(2), pp. 159–168.

Talluri, K. (2004): Revenue Management Under a General Discrete Choice Model of Consumer Behavior. *Management Science* 50(1), pp. 15–33.

Talluri, K. and G. van Ryzin (1998): An Analysis of Bid-Price Controls for Network Revenue Management. *Management Science* 44(11), pp. 1577–1593.

Talluri, K. and G. van Ryzin (1999): A Randomized Linear Programming Method for Computing Network Bid Prices. *Transportation Science* 33(2), pp. 207–216.

Talluri, K. T. and G. J. van Ryzin (2004): *The Theory and Practice of Revenue Management*. Kluwer, Boston (Mass.).

Talluri, K. T. and G. J. van Ryzin (2005): Errata for the book 'The Theory and Practice of Revenue Management'. Online publication as of May 28, 2005, accessed on December 31, 2005, URL http://www.econ.upf.edu/~talluri/RM_book_errata.pdf.

Tirole, J. (1988): *The Theory of Industrial Organization*. MIT Press, Cambridge (Mass.).

Toh, R. S. and P. Raven (2003): Perishable Asset Revenue Management: Integrated Internet Marketing Strategies for the Airlines. *Transportation Journal* 42(4), pp. 30–43.

Triantis, G. G. (2000): Unforeseen Contingencies. Risk Allocation in Contracts. In: B. Bouckaert and G. De Geest (eds.), *The Regulation of Contracts*, vol. 3 of *Encyclopedia of Law and Economics*, chap. 4500, Edward Elgar, Cheltenham, pp. 100–116.

Trigeorgis, L. (1995): *Real Options in Capital Investment – Models, Strategies, and Applications*. Praeger, Westport (Conn.).

Tsay, A. A. (1999): The Quantity Flexibility Contract and Supplier-Customer Incentives. *Management Science* 45(10), pp. 1339–1358.

Tsay, A. A., S. Nahmias, and N. Agrawal (1999): Modeling Supply Chain Contracts: A Review. In: S. Tayur, M. Magazine, and R. Ganeshan (eds.), *Quantitative Models for Supply Chain Management*, chap. 10, Kluwer, Boston (Mass.), pp. 299–336.

Tscheulin, D. K. and J. Lindenmeier (2003): Yield-Management – Ein State-of-the-Art. *Zeitschrift für Betriebswirtschaft* 73(6), pp. 629–662.

Varian, H. A. (1999): *Intermediate Microeconomics: A Modern Approach*. 5th ed., Norton, New York.

Vulcano, G., G. van Ryzin, and C. Maglaras (2002): Optimal Dynamic Auctions for Revenue Management. *Management Science* 48(11), pp. 1388–1407.

Wang, G. H. K., W. Maling, and E. McCarthy (1981): Functional Forms and Aggregate U.S. Domestic Air Cargo Demand: 1950- 1977. *Transportation Research Part A* 15(3), pp. 249–256.

Weatherford, L. R. and S. E. Bodily (1992): A Taxonomy and Research Overview of Perishable-Asset Revenue Management: Yield Management, Overbooking, and Pricing. *Operations Research* 40(5), pp. 831–844.

Weatherford, L. R. and P. E. Pfeifer (1994): The Economic Value of Using Advance Booking of Orders. *Omega* 22(1), pp. 105–111.

Weatherford, L. R. and S. Pölt (2002): Better unconstraining of airline demand data in revenue management systems for improved forecast accuracy and greater revenues. *Journal of Revenue and Pricing Management* 1(3), pp. 234–254.

Weisskopf, G. (1984): *Der Luftfrachtmarkt, unter besonderer Berücksichtigung der Beziehungen zwischen Luftverkehrsgesellschaften und Luftfrachtspediteuren*. Juris, Zürich.

Whang, S. (1995): Coordination in operations: A taxonomy. *Journal of Operations Management* 12(3/4), pp. 413–422.

Whitin, T. M. (1955): Inventory Control and Price Theory. *Management Science* 2(1), pp. 61–68.

Williamson, E. L. (1988): *Comparison of Optimization Techniques for Origin-Destination Seat Inventory Control*. Master thesis, available at url http://hdl.handle.net/1721.1/14572, Massachusetts Institute of Technology, Cambridge (Mass.).

Williamson, O. E. (1979): Transaction-Cost Economics: The Governance of Contractual Relations. *Journal of Law & Economics* 22(2), pp. 233–261.

Winkler, R. L., G. M. Roodman, and R. R. Britney (1972): The Determination of Partial Moments. *Management Science* 19(3), pp. 290–296.

Wu, D. J., P. R. Kleindorfer, and J. E. Zhang (2002): Optimal bidding and contracting strategies for capital-intensive goods. *European Journal of Operational Research* 137(3), pp. 657–676.

Zhang, A. and Y. Zhang (2002): A model of air cargo liberalization: passenger vs. all-cargo carriers. *Transportation Research Part E* 38(3–4), pp. 175–191.

Zhao, W. and Y.-S. Zheng (2001): A Dynamic Model for Airline Seat Allocation with Passenger Diversion No-Shows. *Transportation Science* 35(1), pp. 80–98.

Index

List of Tables

List of Figures

Lecture Notes in Economics and Mathematical Systems

For information about Vols. 1–483
please contact your bookseller or Springer-Verlag

Printing: Krips bv, Meppel
Binding: Stürtz, Würzburg